PHENOTYPIC PLASTICITY

MOLECULAR MECHANISMS, EVOLUTIONARY SIGNIFICANCE AND IMPACT ON SPECIATION

SCIENCE, EVOLUTION AND CREATIONISM

Additional books in this series can be found on Nova's website
under the Series tab.

Additional e-books in this series can be found on Nova's website
under the e-book tab.

GENETICS - RESEARCH AND ISSUES

Additional books in this series can be found on Nova's website
under the Series tab.

Additional e-books in this series can be found on Nova's website
under the e-book tab.

PHENOTYPIC PLASTICITY

MOLECULAR MECHANISMS, EVOLUTIONARY SIGNIFICANCE AND IMPACT ON SPECIATION

JANET B. VALENTINO

AND

PATRICIA C. HARRELSON

EDITORS

New York

NOTICE TO THE READER

Library of Congress Cataloging-in-Publication Data

ISBN: 978-1-62618-404-6
Library of Congress Control Number: 2013934502

Published by Nova Science Publishers, Inc. † New York

CONTENTS

PREFACE

Phenotypic plasticity is the ability of an organism with a given genotype to respond to changing environmental conditions by undergoing a phenotypic adaptation. The authors in this book discuss the molecular mechanisms, evolutionary significance and impact on speciation of phenotypic plasticity. Topics discussed include the role and evolution of adaptive plasticity during the colonization of novel environments; phenotypic plasticity in functional traits of woody species in tropical dry forests; understanding the mechanisms of chemical communication and its consequences on aquatic communities using the freshwater crustacean daphnia model; and phenotypic plasticity of plants in response to environmental change.

Chapter 1   1. Over the last few decades, research on adaptive phenotypic plasticity has bloomed thanks to a deeper empirical and theoretical understanding of genotype x environment interactions. One of the most important advances stemmed from the newly accepted views that adaptive plasticity could evolve like a trait, was subject to natural selection, and was a potential driver for both allopatric and sympatric speciation. In particular, the role of adaptive plasticity in facilitating the colonisation of novel environments has become clearer.

2. The successful colonisation of novel environments was known to be often associated with altered expressions of behaviour, morphology or physiology. However, little was known of the role of plastic responses in the early stages of colonisation, the importance of putative costs of plasticity, the mechanisms by which plastic traits may be genetically assimilated, and the rates at which such transitions occur.

3. Tiger snakes (*Notechis scutatus*) are large venomous elapid snakes broadly distributed across southern and eastern Australia. Between six and ten thousand years ago, rising sea levels fragmented previously-continuous

populations in southern Australia, submerging the broad coastal plain that had linked eastern and western populations. As a result, the gene pool of *Notechis* became fragmented also, with isolated populations in Western Australia, South Australia, Tasmania, and on some former mountaintops, which are now offshore islands. Recent work revealed an extremely low genetic divergence between Tiger snake populations across their range. Nonetheless, most populations of Tiger snakes diverge from each other for a large array of morphological and behavioural traits, and occur in highly variable ecological niches in terms of habitat, diet, or predation pressure. Tiger snake populations are found in wet habitats on the mainland through to highly arid islands, where their typical prey (frogs) do not occur and sea bird chicks form the basis of an alternative diet.

4. Larger prey present on islands has generated powerful selective regimes driving increased body and head size (i.e., swallowing abilities) as well as increased levels of adaptive plasticity in head growth in response to prey size. Young snakes from recently isolated populations (< 6000 years) are able to accelerate head growth, and hence swallowing performance, in response to large prey, while snakes from 'older' islands are born larger but no longer exhibit plasticity in response to prey size. While plasticity in head growth rate may prove advantageous in the early stage of island colonisation, it was proposed that canalised versions of the relevant traits (head growth rates) may eventually outcompete plastic versions as matching prey size via plasticity incurred fitness costs.

5. The calculation of evolutionary rates in snake body size at birth showed that in the early stages of island colonisation directional selection may combine with adaptive plasticity to boost evolutionary rates towards a target phenotype (optimal body size to match prey size). When selection for 'optimal' phenotypes and adaptive plasticity share a common direction towards a target phenotype, both mechanisms may act in synergy and generate accelerated evolution, until plastic phenotypes are eventually outcompeted by canalised ones. Although previously supported by mathematical models, this study provides strong empirical support of the Baldwin accelerating effect: adaptive plasticity can speed up evolution towards an optimal phenotype.

Chapter 2 – The authors analysed the phenotypic plasticity in nineteen functional traits (FTs) (seven morphological and twelve physiological) in tree and shrub species across the five study sites located in a tropical dry forest (TDF), showing variable soil moisture content (SMC). The aim was to observe the range of FTs in tree and shrub species across the study sites. Further, the response of FTs to SMC across species and sites was analysed. The authors also studied the relationships among FTs and between FTs and soil properties across sites. Results showed that the plasticity in FTs significantly varied across the study sites. The plasticity in FTs also differed significantly across species. All FTs under study affect relative growth rate (RGR) of the tree and shrub species directly or indirectly. However, the strength of effect is determined by environmental parameters and in case of TDF soil water availability is the important parameter. Plasticity in FTs due to changes in environmental parameters explained the variations in RGR. Step-wise multiple regression indicates that more than 80% variability in RGR can be explained by canopy cover (CC), leaf area index (LAI), specific leaf area (SLA) and leaf intrinsic water use efficiency (WUEi) alone. First three variables represent quantity of photosynthetic surface and last represent water use economy of a species. All these are also significantly modulated by soil moisture availability. Important point to note here is that photosynthetic rate (A_{max}) is not an important parameter to determine RGR in TDF where water economy and extended period of leaflessness are critical.

Chapter 3 – Phenotypic plasticity is defined as the ability of an organism with a given genotype to respond to changing environmental conditions by undergoing a phenotypic adaptation. Predator-induced defences are one form of phenotypic plasticity. Only when needed i.e. in the presence of a predator, organisms build defensive structures (e.g. crests, thorns or spines) and/or behaviours in a cost-benefit efficient manner. In aquatic ecosystems these phenotypic changes are frequently triggered by predator-specific chemical cues (kairomones). This kind of chemical information transfer has a major impact on food-web interactions as inducible defences dampen predator-prey oscillations. This stabilizes populations and increases their persistence.

The ecology and evolution of these chemically mediated anti-predator defences have been intensely studied especially in the freshwater crustacean *Daphnia*. Together with the sequencing of the genome the freshwater crustacean *Daphnia* has been established as a new model organism.

The *Daphnia* genome holds remarkable aspects that are hypothesized to allow for the observed flexibility towards environmental changes. An enormous number of at least 30,000 genes, several of which are members of recently diverged gene families, characterize the *Daphnia* genome. Such exceptional and almost unique features call for more detailed studies. Besides the study of the genome evolution in *Daphnia*, the determination of the functional genomic and physiological mechanisms that enable *Daphnia* to adapt to changing environmental conditions are of particular interest. These mechanisms include processes suitable for the detection of environmental conditions, and features that allow an effective phenotypic alteration.

This chapter will review information on the cellular and neuronal mechanisms involved in the physiology underlying the formation of inducible defences. This will be discussed together with information available from genome data.

The authors will consider state of the art research and suggest future directions leading to a more detailed understanding of how phenotypes adapt to changing environments.

Chapter 4 – Plants can adjust their phenotype in response to changing environmental conditions through developmental plasticity. The study of phenotypic plasticity is naturally interdisciplinary and encompasses aspects of behaviour, development, ecology, evolution, genetics, genomics, and multiple physiological systems at various levels of biological organization. In this article, the authors provide a brief description of phenotypic plasticity in plants; examine its potential adaptive significance; emphasize recent molecular approaches that provide novel insight into underlying mechanisms, and highlight examples of phenotypic plasticity from plants behaviour in response to various environmental conditions.

In: Phenotypic Plasticity   ISBN: 978-1-62618-404-6
Editors: J. Valentino and P. Harrelson   © 2013 Nova Science Publishers, Inc.

Chapter 1

ROLE AND EVOLUTION OF ADAPTIVE PLASTICITY DURING THE COLONIZATION OF NOVEL ENVIRONMENTS

Fabien Aubret[1,2,]*

[1]Station d'Ecologie Expérimentale du CNRS à
Moulis USR 2936, Moulis, Saint-Girons, France
[2]School of Biological Sciences A08,
University of Sydney, NSW, Australia

ABSTRACT

1. Over the last few decades, research on adaptive phenotypic plasticity has bloomed thanks to a deeper empirical and theoretical understanding of genotype x environment interactions. One of the most important advances stemmed from the newly accepted views that adaptive plasticity could evolve like a trait, was subject to natural selection, and was a potential driver for both allopatric and sympatric speciation. In particular, the role of adaptive plasticity in facilitating the colonisation of novel environments has become clearer.

2. The successful colonisation of novel environments was known to be often associated with altered expressions of behaviour, morphology or

[*] Corresponding author: F. Aubret, Station d'Ecologie Expérimentale du CNRS 09200 Moulis, Saint-Girons, France. E-mail: aubret@dr14.cnrs.fr; Ph: +33 5 61 04 03 60; Fax: + 33 5 61 96 08 51.

physiology. However, little was known of the role of plastic responses in the early stages of colonisation, the importance of putative costs of plasticity, the mechanisms by which plastic traits may be genetically assimilated, and the rates at which such transitions occur.

3. Tiger snakes (*Notechis scutatus*) are large venomous elapid snakes broadly distributed across southern and eastern Australia. Between six and ten thousand years ago, rising sea levels fragmented previously-continuous populations in southern Australia, submerging the broad coastal plain that had linked eastern and western populations. As a result, the gene pool of *Notechis* became fragmented also, with isolated populations in Western Australia, South Australia, Tasmania, and on some former mountaintops, which are now offshore islands. Recent work revealed an extremely low genetic divergence between Tiger snake populations across their range. Nonetheless, most populations of Tiger snakes diverge from each other for a large array of morphological and behavioural traits, and occur in highly variable ecological niches in terms of habitat, diet, or predation pressure. Tiger snake populations are found in wet habitats on the mainland through to highly arid islands, where their typical prey (frogs) do not occur and sea bird chicks form the basis of an alternative diet.

4. Larger prey present on islands has generated powerful selective regimes driving increased body and head size (i.e., swallowing abilities) as well as increased levels of adaptive plasticity in head growth in response to prey size. Young snakes from recently isolated populations (< 6000 years) are able to accelerate head growth, and hence swallowing performance, in response to large prey, while snakes from 'older' islands are born larger but no longer exhibit plasticity in response to prey size. While plasticity in head growth rate may prove advantageous in the early stage of island colonisation, it was proposed that canalised versions of the relevant traits (head growth rates) may eventually outcompete plastic versions as matching prey size via plasticity incurred fitness costs.

5. The calculation of evolutionary rates in snake body size at birth showed that in the early stages of island colonisation directional selection may combine with adaptive plasticity to boost evolutionary rates towards a target phenotype (optimal body size to match prey size). When selection for 'optimal' phenotypes and adaptive plasticity share a common direction towards a target phenotype, both mechanisms may act in synergy and generate accelerated evolution, until plastic phenotypes are eventually outcompeted by canalised ones. Although previously supported by mathematical models, this study provides strong empirical support of the Baldwin accelerating effect: adaptive plasticity can speed up evolution towards an optimal phenotype.

INTRODUCTION

Many speciation events, indeed, entire adaptive radiations, have resulted from the colonisation of novel environments by small isolated populations. This type of speciation is best seen on oceanic islands (Darwin, 1845). The high incidence of genetic divergence between island/mainland populations is a reflection of the (1) conditions (and hence, selective pressures) that often differ tremendously from those experienced by mainland conspecifics; (2) small initial population sizes that facilitate founder effects and rapid shifts in allele frequency; and (3) lack of gene flow with the source area that allows the isolated population to follow a unique evolutionary trajectory. For these reasons, islands have long been prime model systems to explore how adaptive shifts and speciation occur in animal populations. Moreover, invasions often occur independently on several adjacent but isolated islands, thereby offering replicates of simplified ecosystems where it is feasible to identify and quantify divergent selective forces.

While the mechanisms generating genetic divergence (i.e., adaptive radiation) between long isolated and mainland populations are well documented and understood (Grant, 1999; Gravilets and Vose, 2005), deciphering the initial evolutionary stages of island colonisation has been the subject of both passionate debate and tremendous advances over the last few decades. Above all, the potential role played by phenotypic plasticity has been widely recognised.

Phenotypic plasticity is the property of a given genotype to produce different phenotypes in response to distinct environmental conditions (Stearns 1989). The putative adaptive value of phenotypic plasticity has recently become a major focus of theoretical and empirical studies by evolutionary biologists (Scheiner, 1993; Via et al., 1995; Schlichting and Pigliucci, 1998; Pigliucci and Murren, 2003). Plastic responses to environmental cues are often assumed to be adaptations that allow organisms to develop phenotypes appropriate to the environments they experience (Kingsolver, 1995, 1996; Schmitt et al., 1995; Dudley and Schmitt, 1996). For these reasons, the successful colonisation of new environments by animals was often found to be associated with altered behaviours, feeding strategies (including feeding apparatus), defence mechanisms and other forms of plastic responses (Ehrlich, 1989; Holway and Suarez, 1999; West-Eberhard, 2003; Yeh and Price, 2004; Fitzpatrick, 2012). That is, the novel conditions directly induce changes in an individual's behaviour, morphology and physiology.

These plastic responses may allow a population to persist under temporarily stressful conditions (population establishment) and/or allow survival of the population under novel environmental conditions (population persistence), leaving more time for mutation, recombination, and selection to fine-tune the level of adaptation (Baldwin, 1896; Waddington, 1942; West-Eberhard, 2003). Phenotypic plasticity also may extend the ecological range of a species by allowing gradual adaptation to selection pressures otherwise not encountered, and creating the opportunity for natural selection leading to genetic assimilation (Waddington, 1961). In the absence of plastic responses, population growth may be low or negative during the early generations of colonisation, so that extinction may occur before adaptive evolution is possible (Chevin et al., 2010).

Island populations may lead to new species which are, by definition, genetically separated from the ancestors. How, then, does plasticity interact with environmental conditions to produce genetic changes? Indeed, if individuals can attain high fitness in the new environment by displaying plastic responses, then why should there be directional selection (i.e., other than selecting for increasingly plastic individuals) and genetic differentiation from the source at all?

Once the new population is established, and if the new conditions remain consistent from one generation to the next, then standard evolutionary theory predicts the loss of plasticity and the evolution of a canalised phenotype, notably because of the costs associated with plasticity (Behera, 1994; Mayley, 1997; Relyea, 2002; Steinger et al., 2003). Clearly, the two processes interact, so that selection favours any allele that produces a temporarily optimal phenotype, even if that allele acts via norms of reaction (i.e., it produces the "correct" phenotype through developmental plasticity, in response to the conditions in the new environment) rather than via some more direct canalised mechanism (i.e., an allele that generates the "correct" phenotype regardless of the conditions encountered by the organism). These initially plastic genotypes may then be viewed as constituting "bridges" from one generation to the next, allowing time for favourable genetic variants to appear by mutation or recombination and eventually the invasion of canalised phenotypes in the established population. Importantly though, this scenario does not apply in a newly-colonised environment that remains predictably variable (i.e., both the ancestral and novel conditions may exist through time and/or space). In this case, plastic phenotypes would retain higher mean fitness across environments than canalised phenotypes, and plasticity levels should not decrease over time.

Hence, developmental plasticity can generate phenotypic diversity and a corresponding opportunity for selection; and thus can facilitate natural selection and a directional shift in the population mean values for fitness-relevant traits (Waddington, 1961; Schlichting and Pigliucci, 1995; West-Eberhard, 2003). These modern ideas relate plasticity to Waddington's (1942, 1961) and Schmalhausen's (1949) concept of genetic assimilation.

A classic example of this phenomenon is provided by the extensive evolutionary radiation of *Anolis* lizards, both in the Caribbean and in Central and South America. Adaptation to diverse structural habitats, differing in perch dimensions, involved evolutionary alterations in limb length (Losos et al., 2000). Species of *Anolis* lizards that use broad surfaces have longs legs, which provide enhanced maximal sprint speed. However species that use narrow surfaces have short legs, which permit careful movements. Hatchlings of *Anolis sagrei* were raised in terraria provided with only broad or narrow surfaces.

After five months, lizards in the broad treatment had developed relatively longer hindlimbs than lizards in the narrow treatment. By matching phenotype to local conditions, phenotypic plasticity may have played an important role in the evolutionary radiation of *Anolis* lizards across habitat types throughout the diverse islands of the Caribbean (Losos et al., 1997, 1998, 2000).

Once the populations occupied the new habitats, natural selection would have then favoured any new non-plastic genetic variation that enhanced organismal performance in this habitat (i.e., genetic assimilation, Losos et al., 2000).

Although the potential importance of genetic assimilation to evolutionary changes in founder populations has been theoretically demonstrated (Behera, 1994; Rollo, 1994; Mayley, 1996; Schlichting and Pigliucci, 1998; Pigliucci and Murren, 2003; Price et al., 2003; West-Eberhard, 2003; Schlichting, 2004; Badyaev, 2005; Lande 2009), empirical evidence on this topic is rare. Perhaps for this very reason, it was recently suggested that genetic assimilation only played a minor role in evolution (de Jong, 2005). On the other hand, some authors pointed out that genetic assimilation may in fact occur on such short timescales that it is difficult to detect except under unusual circumstances, hence a scarcity of empirical examples (Pigliucci and Murren, 2003; Pigliucci et al., 2006).

Despite these recent advances in the field, little has been done that explicitly incorporates the transitional role of phenotypic plasticity leading to macroevolution level biological change (i.e., genetic assimilation, costs of plasticity, speciation).

That is, empirical evidence is lacking on critical issues such as the putative costs of plasticity, the replacement of plastic traits by canalised versions of the traits (i.e., erosion of plasticity), and the rates at which such transitions occur. In an attempt to empirically answer these questions, a study was conducted on a mosaic of snake populations isolated (or introduced) on islands across southern Australia from less than 30 years ago to over 9000 years ago.

GENERAL METHODS

Study System and Model

Tiger snakes (*Notechis scutatus*) are large venomous elapid snakes widely distributed across southern and eastern Australia. Between six and ten thousand years ago, rising sea levels fragmented previously-continuous populations in southern Australia, submerging the broad coastal plain that had linked eastern and western populations (Rawlinson, 1974; Schawner, 1985). As a result, the gene pool of *Notechis* became fragmented also, with isolated populations in Western Australia, South Australia, Tasmania and on some former mountaintops, which are now offshore islands (Rawlinson, 1974; Keogh et al., 2005). A small number of these populations are the result of recent human introductions (Carnac Island, Trefoil Island; Aubret et al., 2004a; Aubret and Shine, 2009).

Recent work by Keogh et al. (2005) revealed extremely low genetic divergence between Tiger snake populations across their range, to the extent that Tiger snakes are believed to be a single polymorphic species (genetic divergence is at most 1.4% between Western Australia and populations in south-eastern Australia). Within south-eastern Australia for example, the maximum genetic divergence is only 0.38% between populations of island giants, island dwarfs and mainland Tiger snakes (Keogh et al. 2005). Nevertheless, *Notechis* exhibits strong variations in ecological and morphological traits between mainland and island populations and also between adjacent islands (Shine, 1977a, 1987; Schwaner, 1985; Aubret et al., 2004a, b; Bonnet et al., 2005). Detailed studies notably showed inter-population variations in scalation (Aubret et al., 2004a), feeding preferences (Aubret et al., 2006) or defensive behaviour (Bonnet et al., 2005; Aubret et al., 2011), as well as some level of plasticity in locomotor performances (Aubret and Shine, 2007, 2008), or thermoregulatory strategies (Aubret and Michniewicz, 2010; Aubret and Shine, 2010).

The most striking example of trait variation stemmed from the study of body size amongst Tiger snake populations (Keogh et al., 2005). That is, mainland Tiger snakes typically reach sizes of approximately 78-92 cm snout-vent length (SVL) and weigh around 400 g (Shine, 1987; Schwaner and Sarre, 1990). Roxby Island, off South Australia, on the other hand is populated only by dwarfs that reach an average of 70 cm SVL and weigh less than 200 g, while Mount Chappell Island, off Tasmania, is populated by giants that can attain sizes of 160 cm and weigh well over 1kg (Shine, 1987; Schwaner and Sarre, 1988, 1990). Giant Tiger snakes are also found on the islands of the Nuyts Archipelago in the Great Australian Bight, and on Hopkins and Williams Island in the Port Lincoln and Neptune Island groups in South Australia (Schwaner, 1985; Robinson et al., 1996).

Comparative data suggest that these striking variations in body size are linked to prey availability, either by adaptive rigid genetic expressions (i.e., canalised shifts in life history traits such as body size at birth or growth rates) and/or by developmentally plastic responses (i.e., adaptive reaction norm to food availability; Aubret et al., 2004b; Keogh et al., 2005). Field observations highlighted the fact that snakes were larger in places where food resources consisted of larger prey (notably, sea bird chicks) and smaller where the snakes were restricted to smaller prey (skinks; Shine, 1987; Schwaner and Sarre, 1990).

Materials and Methods

Eleven Tiger snake populations were regularly surveyed during numerous field trips from 2001 to 2010 (Figure 1). Study sites included Carnac Island, Herdsman Lake, and Joondalup Lake in Western Australia; Williams Island, Reevesby Island, and Hopkins Island in South Australia; Trefoil Island, Christmas Island, New-Year Island and Swan Island in Tasmania; and mainland Tasmania (Circular Head). A total of 567 snakes were captured by hand, measured in snout-vent length to the nearest ± 0.5 cm and weighed using an electronic scale (± 1 g). All snakes were individually marked by scale clipping for future identification. All were released within 24 hours at their exact site of capture, with the exception of 72 pregnant females that were brought back to the laboratory (University of Western Australia: 2001 to 2003; University of Sydney: 2006 to 2008).

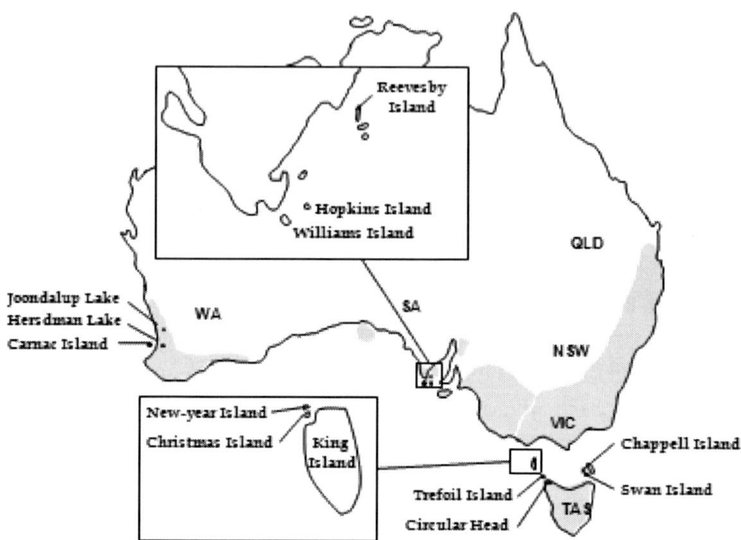

Reproduced from Aubret 2012. The American Naturalist, 179, 756-767. [©] 2012 by The
   University of Chicago.

Figure 1. Distribution map of Australian Tiger snakes showing locations of 12
populations used in this study between 2001 and 2010.

Pregnant females included 21 from Carnac Island, 22 from Herdsman
Lake, 4 from Joondalup Lake, 4 from Williams Island, 4 from Reevesby
Island, 3 from Trefoil Island, 6 from Christmas Island, 4 from New-Year
Island, and 4 from mainland Tasmania.

Females were housed individually in plastic cages in a temperature
controlled room (27°C by day and 20°C by night). All were offered food once
a week (dead mice and chicks) and water *ad libitum*. A lamp provided basking
opportunity.

A total of 1066 neonates were born and measured shortly after birth in
snout-vent length (± 0.1 cm), body mass (± 0.1 g), and sexed by eversion of
the hemipenes. Head size measurements (an indicator of gape size; Aubret and
Shine 2010), were performed using a digital caliper (± 0.01 mm) and included
jaw length, skull length and head width.

Complementary data on snake body size (including 30 adult snakes from
various origins and one complete litter of 18 Reevesby Island snakes) were
gathered from museum collections across Australia as well as from published
literature (Schwaner, 1985; Schwanner and Sarre, 1988, 1990).

Data on prey available to each population were obtained via personal observation as well as from published literature (Schwaner, 1985; Schwanner and Sarre, 1988, 1990; Robinson et al., 1996; Cogger, 2000; Swain and Jones, 2000; Arena and Wooler, 2003; Wilson and Swan, 2003; Aubret et al., 2004a; Chapple, 2005) and museum collections. Prey were measured for body mass (± 0.1 g).

ADAPTIVE PLASTICITY AND ISOLATION TIME

Ecological Context and Experimental Approach

Mainland Tiger snakes typically produce litters consisting of numerous but small neonates (Cogger, 2000). Neonates average 4.48 ± 0.16 g in body mass and 18.49 ± 1.55 cm in snout-vent length (Aubret, 2012). A wide array of small potential prey items is available to young snakes on the mainland, including numerous skinks and frogs species (Cogger, 2000; Aubret, 2012). This ecological context is described here as scenario 1, Figure 2. As sea level started to rise and Tiger snakes became isolated on islands, adults and neonates alike were quickly faced with unfamiliar prey assemblages, consisting of larger and more robust skinks, rats, and a variety of sea-bird chicks (scenario 2, Figure 2; Bonnet et al., 2002; Keogh et al., 2005; Aubret, 2012).

As snakes are gape-limited predators, the size of the prey they are capable of swallowing is dictated by the size of their gape (itself a combination of relative jaw length and absolute head size; King, 2002). Inevitably, neonate snakes with access only to prey too large for them to swallow will starve to death. One can argue that under these circumstances, survival rates of neonate snakes in the early stages of island colonisation would have been low and that a number of newly fledged Tiger snake populations presumably went extinct (Rawlinson, 1974).

In other words, the shift in diet (and above all prey size) inherent to island life generated a huge selective pressure for swallowing performance, hence absolute head size (Aubret et al., 2004b).

It is a reasonable postulation that young snakes that were capable of quickly increasing their head size as a response to prey size survived better (scenario 3, Figure 2) than snakes that were not able to adjust and therefore suffered low (or nil) food intake.

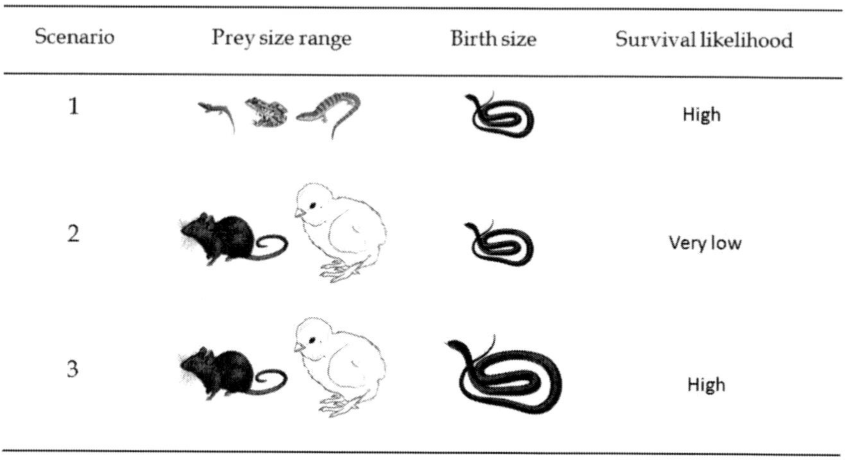

Scenario	Prey size range	Birth size	Survival likelihood
1			High
2			Very low
3			High

Figure 2. Prey assemblages available to mainland (scenario 1) and recently isolated island neonate Tiger snakes (scenario 2 an 3) faced with much larger prey than the staple mainland prey. Survival rates are very low in absence of plasticity for head growth rates but higher when neonates can rapidly adjust their phenotypes, notably head size, as a response to prey size.

As a consequence, it was predicted that (1) snakes from recently isolated populations would display higher levels of plasticity in head growth rates as a response to prey size; (2) snakes from longer isolated systems would display lower levels of plasticity (or no plasticity at all); and (3) snakes from long isolated populations would be born larger (notably in head size) as a consequence of optimal trait canalisation (i.e., genetic assimilation).

In order to test these hypotheses, 191 neonate Tiger snakes were used in a common-garden experiment under the protocol described in Aubret et al. (2004b). The neonates emanated from 31 pregnant females captured on Carnac Island (8 litters); Joondalup Lake (4 litters); Christmas Island (6 litters); New-Year Island (3 litters); Tasmania (4 litters); Trefoil Island (3 litters); and Williams Island (3 litters). For each population, a few neonates were selected from each litter (i.e., split-clutch design, in order to even potential maternal effects across treatments) and allocated to either a "large prey" or "small prey" treatment group.

Each snake was measured in body mass, snout-vent length and head size at birth, and then at two-monthly intervals. Special attention was paid to keep prey for the "large prey" group as close as possible to the upper limit of ingestible prey size for each snake, so that swallowing abilities were consistently challenged throughout the experiment.

Quantifying the Degree of Plasticity

All snakes were fed similar amounts of food over the course of the experiment. The size of prey items differed however, ranging from 0.4 to 11.2 g in the "large prey" group, and from 0.5 g to 3.5 g in the "small prey" group (prey size increased as snakes grew larger; see Aubret and Shine, 2009 for details). After 242 days, phenotypic differences between the "large prey" and "small prey" snakes emerged in some populations but not others (Table 1). For instance, neither young snakes from long isolated Williams Island (9100 BP) nor mainland Joondalup Lake snakes showed a significant difference in jaw growth rates in response to prey size (Figure 3 panel A and panel B respectively). On the other hand, recently introduced Carnac Island snakes (1920's) were capable of sharply increasing head size in response to large prey (Figure 3 panel C). In order to compare the levels of phenotypic plasticity in growth rates amongst populations, the slope of the least-squares regression of Log (relative jaw length) against Log (snake age, i.e., days since birth) was calculated separately for each treatment group in each population (Aubret and Shine, 2009).

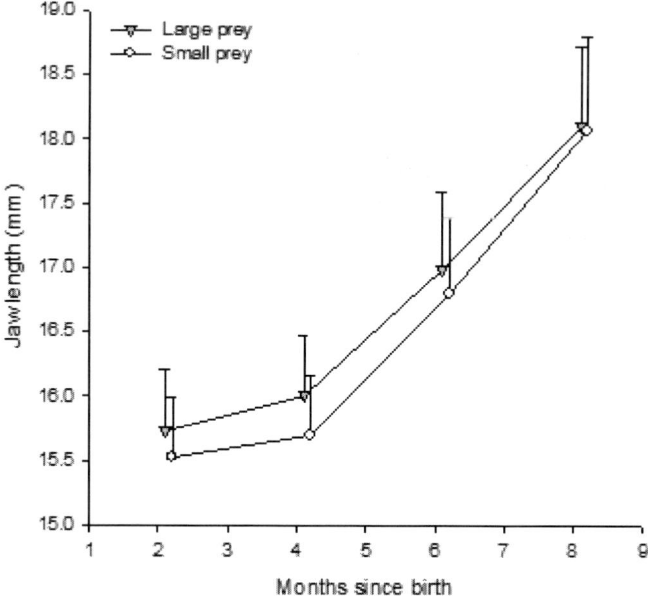

a

Figure 3. (Continued)

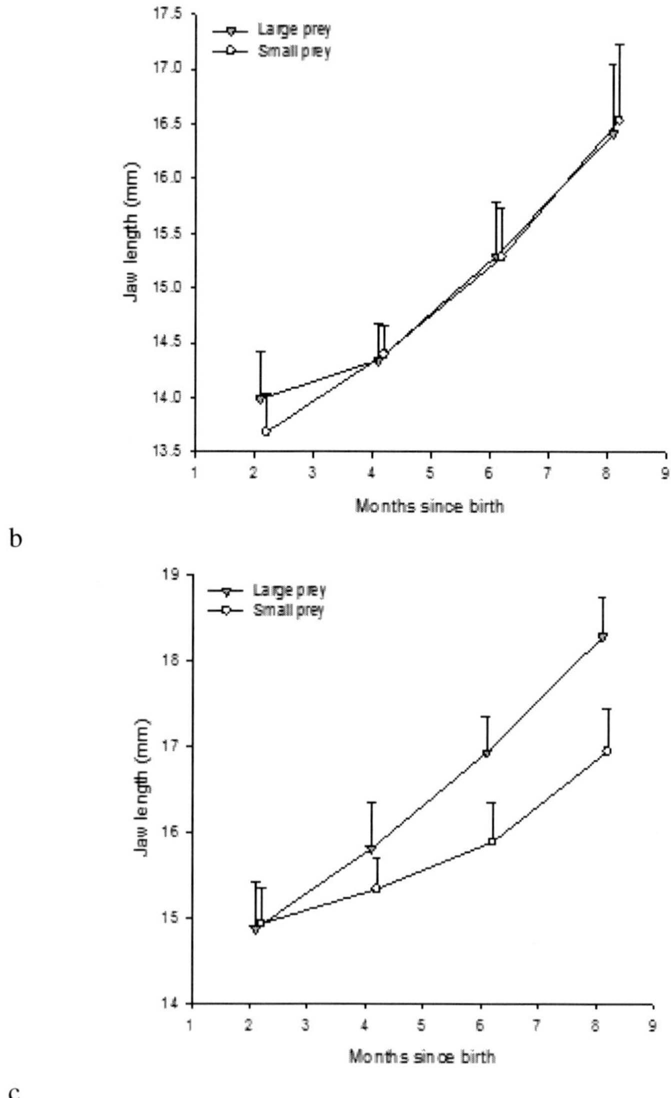

b

c

Figure 3. The degree of adaptive plasticity for jaw growth rates was measured in juvenile Tiger snakes fed either with large (grey triangles) or small (white circles) prey item over a 242 days period. Neither young snakes from long isolated Williams Island (9100 BP) nor mainland Joondalup Lake snakes showed a significant difference in jaw growth rates in response to prey size (Figure 3 panel A and panel B respectively). On the other hand, recently introduced Carnac Island snakes were capable of sharply increasing head size in response to large prey (Figure 3 panel C). Means ± standard deviations are plotted.

Table 1. Results of statistical analyses of data on head measurements of young Tiger snakes (*Notechis scutatus*) raised on either small or large food items. Isolation times for all populations are given (in years before present, BP). Joondalup Lake and Tasmania are source populations. Measurements were taken every two months. Statistical analysis was performed using repeated measures ANOVAs, with treatment as factor, head measurement as the dependent variable through time, and body length as the (changing) covariate. P values are given, and shown in boldface font where P < 0.05. Reproduced from Aubret and Shine 2009 Current Biology 19:1-5

	Williams Island (9100 BP) N = 30	Carnac Island (90 BP) N = 36	New-Year Island (6000 BP) N = 28	Christmas Island (6000 BP) N = 27	Joondalup Lake (Source) N = 25	Tasmania (Source) N = 26	Trefoil Island (40 BP) N = 19	Across populations
Skull length (mm)	0.33	<0.002	<0.006	0.85	0.25	0.17	0.81	<0.032
Jaw length (mm)	0.98	<0.0001	<0.042	0.14	0.11	0.31	<0.013	<0.0001
Head width (mm)	0.60	0.13	0.45	0.43	0.11	0.09	0.73	<0.041

The arithmetic difference in slopes between the "small prey" and "large prey" treatment groups provided an estimate of the levels of plasticity exhibited by each population.

In accordance with predictions 1 and 2, plotting these estimates against isolation time showed that plasticity in relative head size was greatest in Tiger snakes from newly-colonised areas, but lower in snakes from areas that had been colonised or isolated long ago (Least Square Regression N = 5; R = 0.96; $F_{1, 3}$ = 37.54; P < 0.009; Figure 4). In addition, as expected from prediction 3, longer-established populations showed a larger absolute head size and body at birth (Table 2; Figure 5).

This study demonstrated that Tiger snakes from newly-colonised areas had relatively small heads at birth, but were able to increase the rates of growth for head size (hence swallowing performances) when offered large prey items.

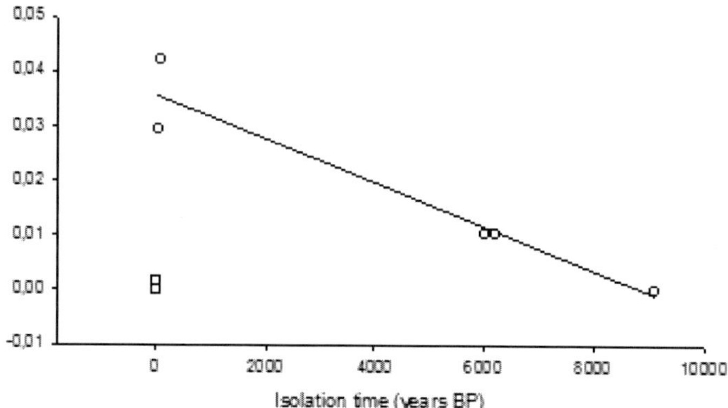

Reproduced from Aubret and Shine 2009. Current Biology 19, 1932-1936. © 2009 Cell Press.

Figure 4. The magnitude of phenotypic plasticity (change in jaw growth rate induced by an experimentally-imposed shift in prey size) declines with the time since island colonisation (isolation time) in Tiger snake populations (Least Square Regression N = 5; R = 0.96; $F_{1,3}$ = 37.54; P < 0.009). Open circles represent island populations and open squares source populations.

Figure 5. Variation in body size and head size at birth amongst populations of Tiger snakes. Shown are neonates from a long isolated Williams island population (top left), a recently isolated Carnac island population (bottom left) and two typical mainland neonate Tiger snakes from Herdsman Lake (centre) and Joondalup lake (top right). All of these neonates were born within a few days of each other and were not fed prior to being photographed.

Table 2. Past, present and future body size in mainland and island Tiger snake populations

Origin *(Number of litters)*	Isolation time (years BP)	Body mass (g) *(target)*	Snout-vent length (cm) *(target)*	Jaw length (mm)	Prey mass (g)
Herdsman L. *(N = 25)*	Mainland	4.74 ± 0.75 *(4.46)*	17.47 ± 1.15 *(18.40)*	13.40 ± 0.51	2.92 ± 0.46
Joondalup L. *(N = 4)*	Mainland	4.35 ± 0.66 *(4.46)*	17.87 ± 0.88 *(18.40)*	13.41 ± 0.14	2.92 ± 0.46
Williams I. *(N = 4)*	9100	8.55 ± 1.23 *(8.92)*	22.24 ± 1.82 *(23.35)*	15.37 ± 0.35	5.42 ± 0.73
Chappell I. *(N = 10)*	9100	7.95 *(7.44)*	23.19 *(21.71)*	-	4.59 ± 0.66
Reevesby I. *(N = 5)*	7700	3.49 ± 0.19 *(3.66)*	16.90 ± 0.83 *(17.51)*	12.58 ± 0.71	2.47 ± 0.39
Christmas I. *(N = 6)*	6000	6.05 ± 0.70 *(10.67)*	20.95 ± 0.84 *(25.29)*	13.59 ± 0.40	6.40 ± 0.81
New-year I. *(N = 4)*	6000	5.31 ± 0.62 *(10.67)*	19.48 ± 1.04 *(25.29)*	13.72 ± 0.59	6.40 ± 0.81
Carnac I. *(N = 22)*	90	5.46 ± 1.09 *(9.39)*	18.64 ± 1.56 *(23.87)*	14.15 ± 0.51	5.68 ± 0.75
Trefoil I. *(N = 3)*	40	4.11 ± 1.17 *(2.98)*	19.75 ± 1.80 *(16.76)*	13.16 ± 0.56	2.09 ± 0.32
Df, F		8, 63; 9.44	8, 64; 9.89	8, 62; 11.85	
P		0.0001	0.0001	0.0001	

In contrast, snakes from long-established populations had larger heads at birth, but displayed little (or no) plasticity in growth rates in response to prey size. These empirical data were in agreement with Waddington's hypothesis of genetic assimilation (Waddington, 1942, 1961; Pigliucci and Murren, 2003; Pigliucci et al., 2006).

As snakes colonised a novel habitat (harbouring unusual prey), the optimal values for major phenotypic traits (in this case, head size and body size) were different from those experienced in the ancestral (mainland) population. This challenge was solved initially by phenotypic plasticity in growth rates.

Remarkably, such plasticity evolved or was selected for very quickly (in under 40 years), providing an empirical illustration of the facilitating role of plasticity in colonising novel environments (West-Eberhard, 2003; Yeh and Price, 2004).

Through time, the initial plasticity was eroded however, and replaced by a canalised version of the optimal phenotypic value (a large head size; Aubret and Shine, 2009).

This observation suggests that genetic assimilation may occur rapidly (over a few decades) even in relatively long-lived, late-maturing animals such as Tiger snakes (2 to 3 years old at maturation; Shine, 1977b; Bonnet et al., 2011). If such rapid change is common, then genetic assimilation will only be observed in studies specifically focussed on the years immediately post-colonisation (Pigliucci and Murren, 2003).

In summary, island Tiger snakes show clear empirical evidence of genetic assimilation, with the elaboration of an adaptive trait shifting from plastic expression through to canalisation within a few thousand years. However, since adaptive plasticity in head growth rates is advantageous in the early stages of colonisation, why do plasticity levels decrease as a function of isolation time? Theoretical work points at potential costs and limits associated with the production of an optimal phenotype by means of plasticity. If adaptive plasticity carries a cost, then is it possible to measure this cost by its impact on organismal fitness?

MEASURING A COST TO ADAPTIVE PLASTICITY

Costs of plasticity are defined as a decrease in fitness even when an optimal phenotype is expressed by plasticity (Pigliucci, 2005). Such costs might include maintenance costs, developmental instability, and/or the time and energy required for the organism to detect the relevant cue (Pigliucci, 2001; Relyea, 2002) and respond adaptively to it via phenotypic modification (De Witt et al., 1998). Apart from putative costs and limits (i.e., most are theoretical but lack empirical evidences; Pigliucci, 2001), adaptive plasticity confers one obvious fitness cost: producing an optimal phenotype using plasticity takes time and energy. Hence, such a phenotype will lag behind a canalised version of an equivalent phenotype in terms of fitness, as long as the selective pressures remain consistent within the environment from one generation to the next (Behera, 1994; Mayley, 1997; De Witt et al., 1998; Pigliucci, 2005; Pigliucci et al., 2006). Even though the concept that adaptive plasticity may confer costs as well as benefits is central to this theory, empirical measurement of such costs has lagged behind conceptual work (but see Krebs and Feder, 1997; Relyea, 2002; Bashey, 2006; Weining et al., 2006; Steiner and Buskirk, 2008).

As previously shown, island Tiger snakes tend to have larger heads than their mainland progenitors (Aubret et al., 2004a, b), but that increase is achieved via different mechanisms on islands that have been recently colonised by snakes, compared to islands where the snakes have been present for many thousands of years (Keogh et al., 2005; Aubret and Shine, 2009). On newly-colonised islands, the increased head size is achieved via adaptive plasticity, with larger prey size stimulating an increase in relative head size (Aubret et al., 2004b).

In contrast, neonates in long-established island populations have larger relative head sizes at birth, and little adaptive plasticity associated with this trait (Aubret and Shine, 2009).

Thus, one way to assess the putative cost of plasticity is to compare the swallowing performances and related growth rates between two groups of snakes: those with large heads from birth, and those that develop large heads via adaptive plasticity in the course of early life (Aubret and Shine, 2010).

In order to avoid potential confounding factors inherent to inter-population comparisons (i.e., other potential adaptations), we experimentally created two phenotypes from a single population, mimicking the early (plastic phenotype) *versus* older (canalised phenotype) stages of a colonisation event (Aubret and Shine, 2009).

We raised sibling neonatal snakes from a recently-colonised island (the highly plastic population of Carnac Island; Aubret and Shine, 2009) either on small or large prey items, to generate two groups of similar-sized snakes with different head sizes and plasticity levels (large head with no plasticity versus small head with plasticity). Carnac Island neonates were born to eight pregnant female Tiger snakes captured on Carnac Island. We created the two phenotypes by raising one group of snakes exclusively on small dead mice (ranging from 1g to 2.6g throughout the experiment) and the others on larger items (1.1g to 11.2 g). Jaw length, snout-vent length and body mass were recorded at regular intervals throughout the 255 days of the experiment.

Once the two phenotypes were created we assessed their swallowing performances by offering the two groups of snakes very large prey items (up to 96% of snake's body mass). If a snake failed to swallow its prey, it was given a small prey item (1.5 to 2 g) the next day to avoid starvation.

The new diet of very large prey thus induced plasticity in head development in the small-headed snakes, but not in the larger-headed animals (i.e., for which the developmental limit of plasticity in head size was reached; Figure 6 panel A).

                                    Fabien Aubret

As expected, large-headed snakes were initially more capable of swallowing large prey items. They did so faster and with fewer jaw protractions than did their smaller-headed siblings (see Aubret and Shine, 2010 for details).

Consequently, large-headed snakes grew more rapidly in body mass and snout-vent length, and were on average 27.6% heavier and 7.9% longer than the small-headed group after 33 days of treatment (Figure 6 panel B).

On the other hand, feeding small-headed Carnac Island snakes with large prey items triggered developmental plasticity in head size: jaw length increased 85.5% faster than in the large-headed group (Aubret and Shine, 2010).

However, producing an equivalent phenotype via plasticity (i.e., jaws as large as those possessed by the large-headed animals at the beginning of the experiment) took approximately 33 days (Figure 6 panel A). This "lag time" can be described as a fitness cost of developmental plasticity.

Under this scenario, the canalised versions of the optimal phenotype may prevail: faster growth may increase survival rate by reducing vulnerability to predation (risk of predation in reptiles is size-dependent; Ferguson and Fox, 1984; Forsman, 1993; Webb and Whiting, 2005) and increasing feeding success (Forsman, 1996; King, 2002; Vincent and Mori, 2008).

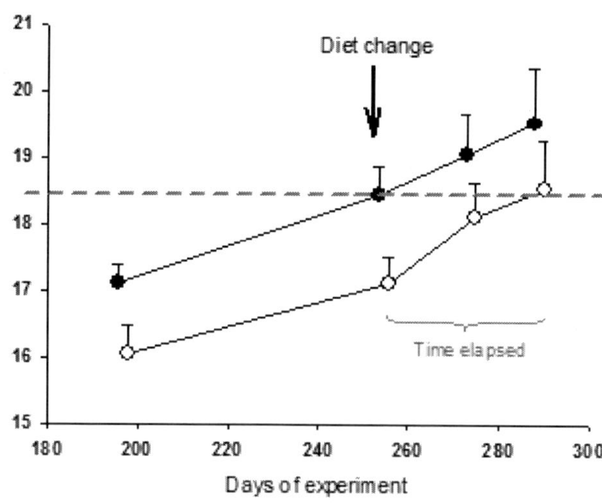

a

Figure 6. (Continued)

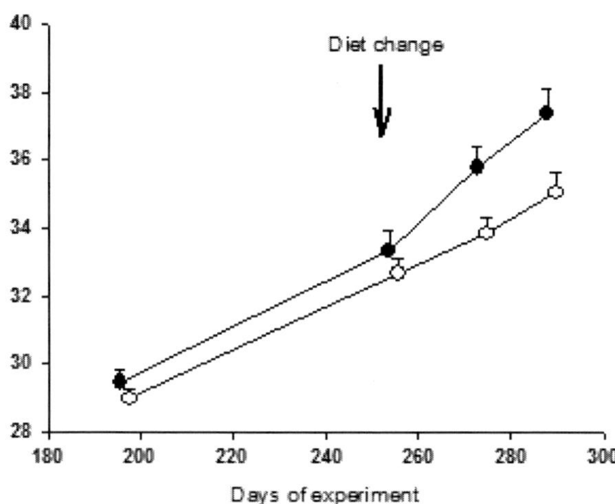

b

Reproduced from Aubret and Shine 2010. The Journal of Experimental Biology 213, 735-739. © 2010 The Company of Biologists Ltd.

Figure 6. Growth rates of two groups of Carnac Island Tiger snakes in jaw size (panel A and snout vent length (panel B) during a 35 day period when they were offered large prey items. One group (open circles) was previously raised on small prey, and thus had relatively small heads that increased in response to the increased prey size (panel A). The other group had been raised on large prey (black circles) and thus had larger heads initially (panel A). Large headed snakes were better able to swallow large prey and grew faster in snout-vent length than the initially small-headed group (panel B). Mean values ± standard errors are plotted.

Additionally, there is a strong positive feedback involved: the larger a snake grows, the larger the prey items that it can ingest, and faster-growing snakes are likely to attain sexual maturity earlier (Ford and Seigel, 1989; Beaupre et al., 1998; Rivas and Burghardt, 2001). This ecological context along with experimental results supports a progressive replacement of plasticity in head growth by large head size at birth along the colonisation timeframe in Tiger snake populations (Aubret and Shine, 2009).

Adaptive plasticity may provide a selective advantage in the early stages of the colonising event by breaching the gap between an ancestral and a new adaptive peak. However, a consistent selective force for larger head size (i.e., where costs outweigh the benefits of plasticity) ultimately will result in the replacement of the plasticity-based pathway with a canalised version of the trait (Pigliucci and Murren, 2003; Pigliucci et al., 2006; Aubret et al., 2009; Lande, 2009). How quickly does this happen?

BODY SIZE EVOLUTION RATES AND ADAPTIVE PLASTICITY

While the role of adaptive plasticity as an initiator for evolutionary change is well established, the idea that adaptive plasticity may also accelerate rates of evolution remains much debated. As shown before, young snakes from recently isolated populations (i.e., < 6000 years) were able to accelerate head growth, and hence swallowing performance, in response to large prey (Aubret and Shine, 2009), snakes from 'older' islands were born larger but no longer exhibited plasticity in response to prey size (Waddington, 1942, 1953; Pigliucci et al., 2006). It was proposed that canalised versions of the relevant traits (head growth rates) may have eventually outcompeted plastic versions as matching prey size via plasticity incurs fitness costs (delay in body size growth and sexual maturity) (Aubret and Shine, 2010). Consequently, birth body size recorded in recently isolated populations (< 6000 years) may not reflect the optimal phenotype.

Morphological data showed that, as amongst adults, body size was highly variable amongst neonates across mainland and island populations of Tiger snakes. Williams Island neonates were for instance twice as heavy and 1.3 times longer at birth than mainland neonates (Table 1). Concurrently, prey distribution also varied tremendously from western to eastern Australia, between mainland and island populations, and also amongst islands (Table 3 and 4).

An average prey body mass was calculated as the arithmetic means between all prey types present at each given site (Table 4). Spearman Rank Order correlations showed a significant correlation between prey body mass and neonate body mass in long-isolated and mainland populations of Tiger snakes ($N = 6$; $R = 0.90$; $t = 4.10$; $P < 0.015$; Figure 7). These results pinpoint prey size as the main driver for the evolution of body size at birth in gape limited predators. On the other hand, data points from more recently isolated populations (i.e., Carnac, Trefoil, Christmas and New-Year Islands) were outliers, suggesting that within these populations, snake body size at birth stood sightly off the optimal match to prey size.

This situation offered an ideal opportunity to estimate evolutionary rates of body size evolution in island Tiger snakes as a function of isolation time, and to reveal if selection towards optimal birth size and adaptive plasticity in growth rates interact following geographic isolation.

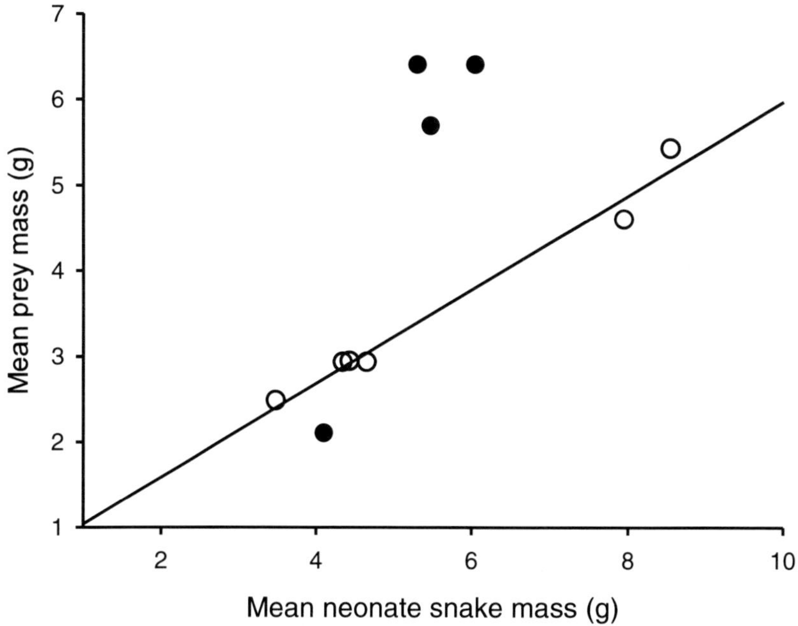

Figure 7. Spearman Rank Order correlations between neonate body mass and prey body mass in long isolated and mainland populations of Tiger snakes (open circles - N = 6; R = 0.90; t = 4.10; P < 0.015). Plots of 4 more recently isolated populations (black circles) are not centred on the regression lines, suggesting that within these populations, selection has not finetuned neonate snake body size to prey size.

Estimating Past, Present and Future Body Size

Field as well as museum data (Aubret, 2012) showed that average prey size and snake birth body size were closely correlated in three long isolated island populations and three mainland populations (Figure 7). Derived equations (Linear Regression of Log [birth snake body mass] against Log [average prey mass]; N = 6; R = 0.99; $F_{1, 4}$ = 213.69; P < 0.00013; regression line equation y = -0.7529+1.7851x; and Log [birth snake snout-vent length] against Log [average prey mass]; N = 6; R = 0.89; $F_{1, 4}$ = 14.81; P < 0.018; regression line equation y = 12.6098 + 1.9819x) and mean available prey sizes were used to calculate an estimate of the target (optimal) phenotype for body mass and snout-vent length in five more recently isolated populations (Table 1).

Table 3. Prey available to neonate snakes across populations

Sites	Prey available
Carnac Island	*Mus musculus, Morethia obscura, Christinus marmoratus*
Herdsman Lake And Joondalup Lake	*M. musculus, Morethia obscura, Ctenotus fallens, Cryptoblepharus plagiocephalus, Hemiergis quadrilineata, C. marmoratus, Lerista christinae, L. distinguenda, Menetia greyii, Crinia insignifera, C. glauerti, Litoria adelaidensis, Limnodynastes dorsalis*
Williams Island	*Egernia multiscutata, Hemiergis peronii, Menetia greyii,C. marmoratus*
Reevesby Island	*Aprasia inaurita, Hemiergis p., Lerista dorsalis, Lerista edwardsae, Morethia adelaidensis, M. obscura, C. marmoratus, Pseudemoia entrecasteauxii, Ctenotus orientalis*
Christmas Island and	*M. musculus, Egernia whitii, Niveoscincus metallicus*
New-year Island	
Tasmania	*M. musculus, E. whitii, Bassiana duperreyi, Lampropholis delicata, Niveoscincus orocrypta, N. mettalicus, N. occelatum, N. pretosius; P. entrecasteauxii, Litoria ewingii, Crinia tasmaniensis, Crinia signifera, Pseudophryne semimarmorata, Limnodynastes tasmaniensis, L. dumerili, Geocrinia laevis*
Trefoil Island	*N. metallicus*
Chappell Island	*M. musculus, E. whitii, N. metallicus, N. ocellatum, B. duperreyi, Lerista bougainvilli*

Reproduced from Aubret 2012. The American Naturalist, 179, 756-767. [©] 2012 by The University of Chicago.

Evolutionary Rates in Body Size, Isolation Time and Plasticity Levels

Mainland Tiger snakes at birth average 4.48 ± 0.16 g in body mass and 18.49 ± 1.55 cm in snout-vent length (Table 1). Evolutionary rates in body mass and snout-vent length per generation (known to be two years on average in Tiger snakes: Shine, 1977b; Bonnet et al., 2011) were calculated as follow: evolutionary rate = (current body size − ancestral body size) / (isolation time / 2).

Evolutionary rates in body mass and snout-vent length per generation were both significantly correlated with isolation time (Linear Regressions; P < 0.0011 and P < 0.0019 respectively; Figure 8 panels A and B).

Table 4. Prey type and size available to neonate snakes across populations (means ± SD are given)

Latin name	Common name	Body mass (g)	N
Aprasia inaurita	Mallee worm-lizard	1.38 ± 0.41	20
Bassiana duperreyi	Eastern three-lined skink	2.65 ± 1.16	11
Bryobatrachus nimbus	Moss froglet	1.02 ± 0.44	8
Christinus marmoratus	Marbled gecko	3.02 ± 0.47	20
Crinia glauerti	Glauert's froglet	1.00	*
Crinia insignifera	Sign-bearing froglet	1.01 ± 0.25	20
Crinia signifera	Common eastern froglet	1.15 ± 0.60	11
Crinia tasmaniensis	Tasmanian froglet	1.42 ± 0.88	10
Cryptoblepharus plagiocephalus	Péron's snake-eyed skink	2.00	*
Ctenotus fallens	West coast laterite skink	8.86 ± 4.84	37
Ctenotus orientalis	Eastern striped skink	6.17 ± 1.79	20
Egernia multiscutata	Southern sand-skink	15.42 ± 4.93	5
Egernia whitii	White's skink	5.10 ± 4.21	6
Geocrinia laevis	Smooth froglet	2.13 ± 0.67	6
Hemiergis quadrilineata	Two-toed earless skink	2.40	*
Hemiergis peronii	Four-toed mulch skink	2.57 ± 0.80	20
Lampropholis delicata	Garden skink	1.02 ± 0.39	4
Lerista bougainvillii	Bougainville's skink	1.50	1
Lerista christinae	Bold-striped slider	1.00	*
Lerista distinguenda	South-western orange-tailed slider	1.15	*
Lerista dorsalis	Southern slider	1.15 ± 0.25	20
Lerista edwardsae	Edwards' slider	1.25	*
Limnodynastes dorsalis	Western banjo frog	2.00	*
Limnodynastes dumerili	Eastern banjo frog	1.14 ± 0.34	9
Limnodynastes tasmaniensis	Spotted marsh frog	2.78 ±1.77	13
Litoria adelaidensis	Slender tree frog	0.85 ± 0.58	32
Litoria burrowsi	Tasmanian tree frog	0.77 ± 0.40	3
Litoria ewingii	Brown tree frog	2.16 ± 0.90	14
Menetia greyii	Common dwarf skink	0.66 ± 0.24	20
Morethia adelaidensis	Samphire skink	2.00	*
Morethia obscura	Obscure skink	2.02 ± 0.34	20
Mus musculus	Mouse	12.00	*
Niveoscincus greeni	Northern snow skink	3.26 ± 1.14	11
Niveoscincus metallicus	Metallic skink	2.09 ± 1.19	14
Niveoscincus microlepidota	Southern snow skink	2.76 ± 1.11	14
Niveoscincus occelatum	Spotted skink	4.21 ± 2.22	13
Niveoscincus orocrypta	Heath cool-skink	2.81 ± 1.14	13
Niveoscincus pretosius	Tasmanian tree skink	1.94 ± 0.66	12
Pseudemoia entrecasteauxii	Southern grass skink	2.66 ± 0.97	20
Pseudophryne semimarmorata	Southern toadlet	1.67 ± 0.56	11

* Data gathered from literature.

Modified from Aubret 2012. The American Naturalist 179: 756-767.[©] 2012 by The University of Chicago.

Level of plasticity for head growth in response to prey size calculated experimentally for 5 island populations in a former study (Trefoil, Carnac, Christmas, New-year and Williams Islands; Aubret and Shine, 2009) were significantly correlated with rates of body mass evolution (Linear Regression; R = 0.89; $F_{1,3}$ = 11.86; P < 0.041; Figure 9), but not snout-vent length (Linear Regression; R = 0 .67; $F_{1,3}$ = 2.40; P < 0.22).

Unusually sized prey can generate an extremely powerful selective pressure on island newcomers (Price et al., 2003; Borenstein et al., 2006; Yeh and Price, 2009), as swallowing success is a *sine qua non* condition to neonate snake survival (Aubret and Shine, 2010; Aubret, 2012).

A likely scenario is that the need to match neonate body size to the larger prey size generates an evolutionary bottleneck, in a first and fast evolutionary stage where the most plastic of the largest individuals are selected (Figure 10).

The fitness advantages of simultaneously possessing both attributes (large body size as well as high levels of plasticity) add up (Figure 10 panel C), and are most likely higher than the advantages of possessing only one or the other trait (Figure 10 panels A and B).

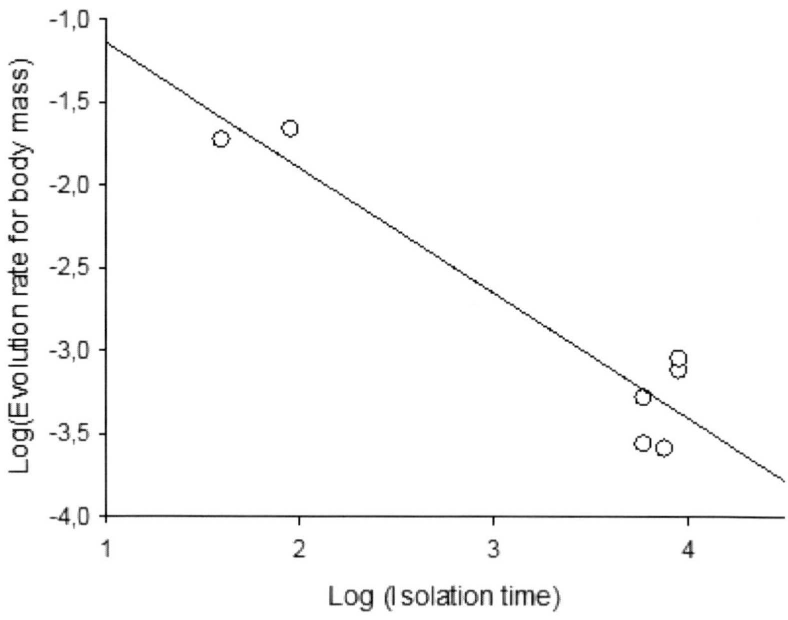

a

Figure 8. (Continued)

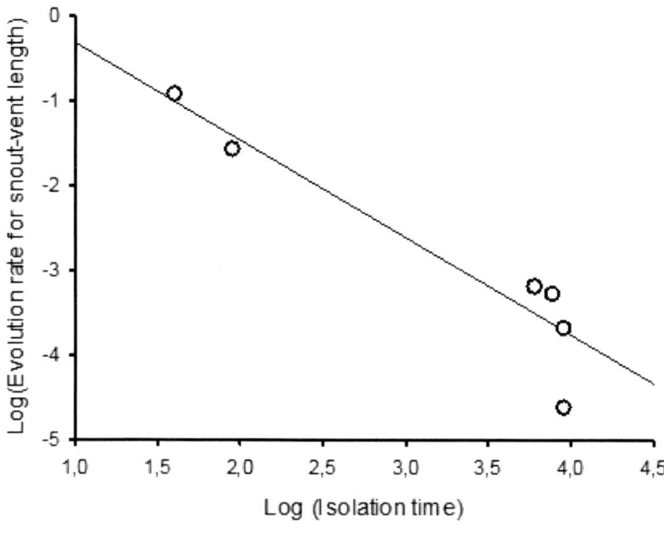

b

Figure 8. Body size evolution rates are plotted as a function of isolation time. Linear Regressions of (1) body mass evolutionary rate per generation against isolation time yielded $R = 0.94$; $F_{1,5} = 44.41$; $P < 0.0011$; (panel A) and (2) snout-vent length evolutionary rates per generation against isolation time yielded $R = 0.94$; $F_{1,5} = 35.56$; $P < 0.0019$ (panel B).

Swallowing performances and body size are indeed linked by a positive feed-back loop that accelerates growth towards sexual maturity (Aubret and Shine, 2007, 2010).

Consequently, such individuals may prevail in the colonising population. As the phenotype nears the target however, fitness costs and limits of plasticity may start to lower fitness (Pigliucci et al., 2006; Aubret and Shine, 2010), interrupting the feed-back loop, thus slowing down evolution (Figure 1).

Eventually, plastic phenotypes may be out competed by canalised phenotypes (Pigliucci et al., 2006; Aubret and Shine, 2010).

CONCLUSION

Although previously supported by mathematical models (Belew and Mitchell, 1996), this study provides the first empirical support of the Baldwin accelerating effect (Baldwin, 1896; Ancel, 2000): adaptive plasticity can speed up evolution towards an optimal phenotype.

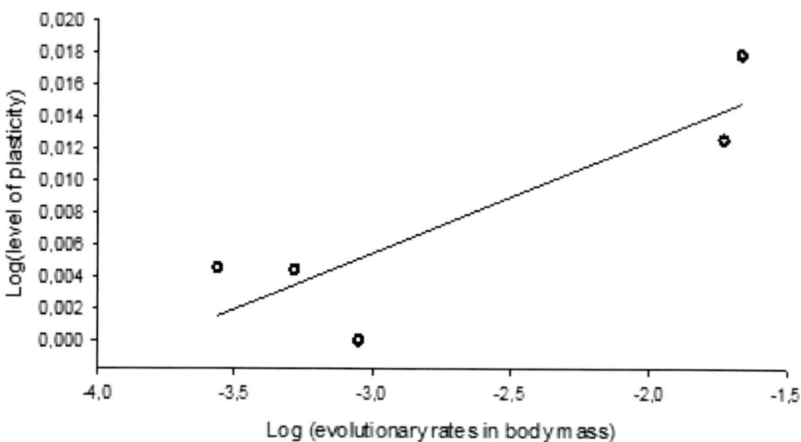

Figure 9. Levels of plasticity for head growth are plotted as a function of isolation time. Level of plasticity were significantly correlated with rates of evolution in body mass (Linear Regression of Log [body mass evolutionary rates] against Log [level of plasticity], $R = 0.89$; $F_{1,3} = 11.78$; $P < 0.042$).

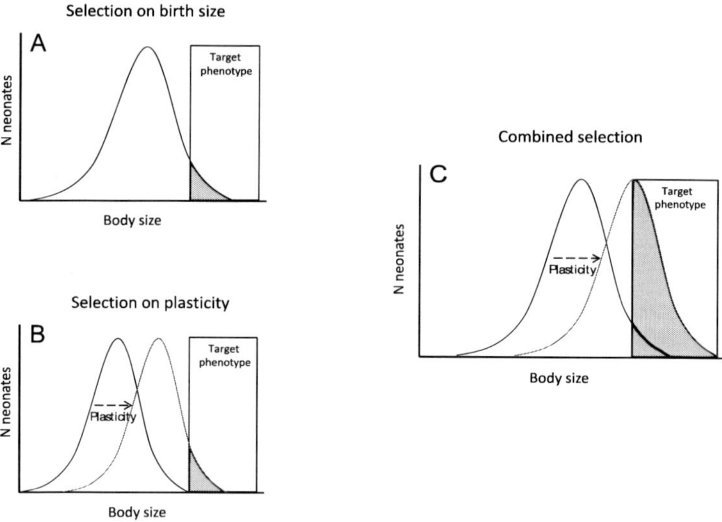

Figure 10. Putative selection regimes on snake neonate body size are presented upon three alternative scenarios in the early stages of island colonization. The distribution of snake neonate body size is plotted. Selection towards a target phenotype acts on neonate body size (panel A) and neonate's levels of plasticity in growth rates (panel B). The combined selective regimes may result in accelerated evolution towards the target phenotype (panel C).

These new insights empirically demonstrate that when selection for 'optimal' phenotypes and adaptive plasticity share a common direction towards a target phenotype, both mechanisms may act in synergy and generate accelerated evolution, until plastic phenotypes are outcompeted by canalized ones.

ACKNOWLEDGMENTS

I thank the Department of Conservation and Land Management (Western Australia; Permit #SF004604 and #CE000347; #SF005274 and #CE001216), the Department of Environment and Heritage (South Australia; permit # M25082 1); the Department of Primary Industries, Water and Environment (Tasmania; Permit # FA 07282) for the issuing of licenses and for their ongoing support of the study. The Animal Ethics Committees of the University of Western Australia (Project # 01/100/177) and the University of Sydney (Project L04/3-2006/4297) approved all procedures. Funding was provided by the University of Western Australia, the Région Poitou-Charentes, the Australian Research Council (ARC), and the Centre National de la Recherche Scientifique (CNRS). I also thank W. Gigg, D. Roberts, D. Bradshaw, X. Bonnet, D. Pearson, R. Shine, J. Thomas, M. Elphick, R. Michniewicz, as well as numerous helpers throughout the years of collection of these data. Léa Lapin assisted in the maintenance of captive snakes. This work is part of the "Laboratoire d'Excellence (LABEX) entitled TULIP (ANR -10-LABX-41).

REFERENCES

Ancel, L. W. (2000). Undermining the Baldwin expediting effect: does phenotypic plasticity accelerate evolution? *Theoretical Population Biology,* 58, 307-319.

Arena, P. C. and Wooller, R. D. (2003). The reproduction and diet of *Egernia kingii* (Reptilia: Scincidae) on Penguin Island, Western Australia. *Australian Journal of Zoology*, 51, 495-504.

Aubret, F. (2012). Body size evolution on islands: are adult size variations in tiger snakes a non-adaptive consequence of selection on birth size? *American Naturalist*, 179, 756-767.

Aubret, F. and Shine, R. (2007). Rapid prey-induced shift in body size in an isolated snake population (*Notechis scutatus*, Elapidae). *Austral. Ecology*, 32, 889-899.

Aubret, F. and Shine, R. (2008). Early experience influences both habitat choice and locomotor performance in tiger snakes. *American Naturalist*, 171, 524-531.

Aubret, F. and Shine, R. (2009). Genetic assimilation and the post-colonization erosion of phenotypic plasticity in island tiger snakes. *Current Biology*, 19, 1932-1936.

Aubret, F. and Michniewicz, R. J. (2010). Warming up for cold water: influence of habitat type on thermoregulatory tactics in a semi-aquatic snake. *Amphibia Reptilia*, 31, 525-531.

Aubret, F. and Shine, R. (2010). Fitness costs may explain the post-colonisation erosion of phenotypic plasticity. *Journal of Experimental Biology,* 213, 735-739.

Aubret, F., Bonnet, X., Maumelat, S., Bradshaw, D., and Schwaner, T. (2004a). Diet divergence, jaw size and scale counts in two neighbouring populations of tiger snakes (*Notechis scutatus*). *Amphibia Reptilia*, 25, 9-17.

Aubret, F., Shine, R. and Bonnet, X. (2004b). Adaptive developmental plasticity in snakes. *Nature*, 431, 261-262.

Aubret, F., Burghardt, G. M., Maumelat, S., Bonnet, X., and Bradshaw, D. (2006). Feeding preferences in two disjunct populations of tiger snakes, *Notechis scutatus*. *Austral. Ecology*, 17, 716-725.

Aubret, F., Michniewicz, R. J. and Shine, R. (2011). Correlated geographic variation in predation risk and antipredator behaviour within a wide-ranging snake species (*Notechis scutatus*, Elapidae). *Austral. Ecology*, 36, 446-452.

Badyaev, A. V. (2005). Stress-induced variation in evolution: from behavioural plasticity to genetic assimilation. *Proceedings of the Royal Society of London* B, 272, 877-886.

Baldwin, J. M. (1896). A new factor in evolution. *American Naturalist*, 30, 441-451.

Bashey, F. (2006). Cross-generational environmental effects and the evolution of offspring size in the Trinidadian guppy *Poecilia reticulata. Evolution*, 60, 348-361.

Beaupre, S. J., Duvall, D. and O'Leile, J. (1998). Ontogenetic variation in growth and sexual size dimorphism in a central Arizona population of the western diamondback rattlesnake (*Crotalus atrox*). *Copeia*, 1998, 40-47.

Behera, N. (1994). Phenotypic plasticity and genetic assimilation in development and evolution. *Bionature*, 14, 1-22.

Belew, R. K. and Mitchell, M. (1996). *Adaptive individuals in evolving populations: models and algorithms.* Reading, MA: Addison-Wesley.

Bonnet, X., Pearson, D., Ladyman, M., Lourdais, L., Bradshaw, S. D. (2002): "Heaven" for serpents? A markrecapture study of Tiger Snakes (*Notechis scutatus*) on Carnac Island, Western Australia. *Austral. Ecology*, 27, 442-450.

Bonnet, X., Aubret, F., Lourdais, O., Ladyman, M., Bradshaw, D., and Maumelat, S. (2005). Do "quiet" places make animals placid? Island versus mainland tiger snakes. *Ethology*, 111, 573-592.

Bonnet, X., Lorioux, S., Pearson, D., Aubret, F., Bradshaw, D., Delmas, V. and Fauvel, T. (2011). Which proximate factor determines sexual size dimorphism in tiger snakes? *Biological Journal of the Linnean Society*, 103, 668-680.

Chevin, L. M, Lande, R. and Mace, G. M. (2010). Adaptation, Plasticity, and Extinction in a Changing Environment: Towards a Predictive Theory. *PLoS Biology* 8(4),e1000357.

Chapple, D. G. (2005). Life-history and reproductive ecology of White's Skink, *Egernia whitii. Australian Journal of Zoology*, 53, 353-360.

Cogger, H. G. (2000). *Reptiles and amphibians of Australia* (6th ed.). Sydney, Australia: Reed New Holland.

Darwin, C. (1845). Journal of researches into the natural history and geology of the countries visited during the voyage of H.M.S. *Beagle round the world, under the Command of Capt. Fitz Roy, R. A.* (2nd. ed.). London, UK: John Murray.

De Jong, G. (2005). Evolution of phenotypic plasticity: patterns of plasticity and the emergence of ecotypes. *New Phytologist*, 166, 101-118.

Dewitt, T. J., Sih, A. and Wilson, D. S. (1998). Costs and limits of phenotypic plasticity. *Trends in Ecology and Evolution*, 13, 77-81.

Dudley, S. A. and Schmitt, J. (1996). Testing the adaptive plasticity hypothesis: Density-dependent selection on manipulated stem length in *Impatiens capensis. The American Naturalist*, 147, 445-465.

Ehrlich, P. R. (1989). Attributes of invaders and the invading processes: vertebrates. In: J. A. Drake, H. A. Mooney, F. di Castri, R. H. Groves, F. J. Kruger, M. Rejmánek, and M. Williamson (Eds.), *Biological invasions: a global perspective* (pp. 315–328). Chichester, UK: John Wiley and Sons.

Ferguson, G. W. and Fox, S. F. (1984). Annual variation of survival advantage of large juvenile side-blotched lizards, *Uta stansburiana*: its causes and evolutionary significance. *Evolution* 38, 342-349.

Fitzpatrick, B. M. (2012). Underappreciated consequences of phenotypic plasticity for ecological speciation. *International Journal of Ecology* 2012: Article ID 256017, doi:10.1155/2012/256017.

Ford, N. B. and Seigel, R. A. (1989). Relationships among body size, clutch size, and egg size in three species of oviparous snakes. *Herpetologica*, 45, 75-83.

Forsman, A. (1993). Survival in relation to body size and growth rate in the adder, *Vipera berus*. *Journal of Animal Ecology*, 62, 647-655.

Forsman, A. (1996). Body size and net energy gain in gape-limited predators: A model. *Journal of Herpetology*, 30, 307-319.

Grant, P. R. (1999). *The ecology and evolution of Darwin's Finches*. Princeton University Press, Princeton, NJ.

Gavrilets, S. and Vose, A. (2005). Dynamic patterns of adaptive radiation. *Proceedings of the National Academy of Science of the US*, 102, 18040-18045.

Holway, D. A. and Suarez, A. V. (1999). Animal behavior: an essential component of invasion biology. *Trends in Ecology and Evolution*, 14, 328-330.

Keogh, J. S., Scott, I. A. and Hayes, C. (2005). Rapid and repeated origin of insular gigantism and dwarfism in Australian tiger snakes. *Evolution*, 59, 226-233.

King, R. B. (2002). Predicted and observed maximum prey size – snake size allometry. *Functional Ecology*, 16, 766-772.

Kingsolver, J. G. (1995). Fitness consequences of seasonal polyphenism in western white butterflies. *Evolution*, 49, 942-954.

Kingsolver, J. G. (1996). Experimental manipulation of wing pigment pattern and survival in western white butterflies. *the American Naturalist*, 14, 7296-7306.

Krebs, R. A. and Feder, M. E. (1997). Natural variation in the expression of the heatshock protein HSP70 in a population of *Drosophila melanogaster* and its correlation with tolerance of ecologically relevant thermal stress. *Evolution* 51, 173-179.

Lande, R. (2009). Adaptation to an extraordinary environment by evolution of phenotypic plasticity and genetic assimilation. *Journal of Evolutionary Biology*, 22, 1435-1446.

Losos, J. B., Warheit, K. I. and Schoener, T. W. (1997). Adaptive differentiation following experimental island colonization in *Anolis* lizards. *Nature*, 387, 70-73.

Losos, J. B., Jackman, T. R., Larson A., de Queiroz, K., and Rodriguez-Schettino, L, (1998). Contingency and determinism in replicated adaptive radiations of island lizards. *Science*, 279, 2115–2118.

Losos, J. B., Creer, D. A., Glossip, D., Goellner, R., Hampton, A., Roberts, G., Haskell, N., Taylor, P. and Etling, J. (2000). Evolutionary implications of phenotypic plasticity in the hindlimb of the lizard *Anolis sagrei*. *Evolution*, 54, 301-305.

Mayley, G. (1996). Landscapes, learning costs, and genetic assimilation. *Evolutionary Computation*, 4, 213-234.

Mayley, G. (1997). Landscapes, learning costs, and genetic assimilation. *Evolutionary Computation* 4, 213-234.

Pigliucci, M. (2001). *Phenotypic plasticity: beyond nature and nurture.* Baltimore, MD: The Johns Hopkins University Press.

Pigliucci, M. and Murren, C. J. (2003). Perspective: genetic assimilation and a possible evolutionary paradox: can macroevolution sometimes be so fast as to pass us by? *Evolution*, 57, 1455-1464.

Pigliucci, M., Murren, C. J. and Schlichting, C. D. (2006). Phenotypic plasticity and evolution by genetic assimilation. *Journal of Experimental Biology,* 209, 2362-2367.

Price, T. D., Qvarnström, A. and Irwin, D. E. (2003). The role of phenotypic plasticity in driving genetic evolution. Proceedings of the Royal Society of London. *Series* B, 270, 1433-1440.

Rawlinson, P. A. (1974). Biogeography and ecology of the reptiles of Tasmania and the Bass Strait area. In: Pages, E. D. Williams (Ed.), *Biogeography and Ecology in Tasmania* (pp. 291-338). The Hague, Netherlands: W. Junk.

Relyea, R. A. (2002). Costs of phenotypic plasticity. *American Naturalist,* 159, 272-282.

Rivas, J. A. and Burghardt, G. M. (2001). Understanding sexual size dimorphism in snakes: wearing the snake's shoes. *Animal Behaviour,* 62, F1-F6.

Robinson, T., Canty, P., Mooney, T., and Ruddock, P. (1996). *South Australia's Offshore Islands.* Canberra, Australia: Australian Heritage Commission.

Rollo, C. D. (1994). *Phenotypes: Their Epigenetics, Ecology and Evolution.* London: Chapman and Hall.

Scheiner, S. M. (1993). Genetics and evolution of phenotypic plasticity. *Annual Review of Ecology and Systematics*, 24, 35–68. *

Schlichting, C. D. (2004). The role of phenotypic plasticity in diversification. In: *Phenotypic Plasticity: Functional and Conceptual Approaches* (ed. T. J. DeWitt and S. M. Scheiner), pp. 191-200. Oxford: Oxford University Press.

Schlichting, C. D. and Pigliucci, M. (1995). Gene regulation, quantitative genetics, and the evolution of reaction norms. *Evolutionary Ecology* 9, 154-168

Schlichting, C. D. and Pigliucci, M. (1998). *Phenotypic Evolution: A Reaction Norm Perspective*. Sunderland, MA: Sinauer Associates.

Schmalhausen, I. I. (1949). *Factors of Evolution*. Philadelphia, PA: Blakiston.

Schmitt, J., McCormac, A. C. and Smith, H. (1995). A test of the adaptive plasticity hypothesis using transgenic and mutant plants disabled in phytochrome-mediated elongation responses to neighbors. *The American Naturalist,* 14, 6937-953.

Schwaner, T. D. (1985). Population structure of black tiger snakes, *Notechis ater niger*, on off-shore islands of South Australia. In: G. Grigg, R. Shine, and H. Ehmann (Eds.), *Biology of Australasian frogs and reptiles* (pp. 35-46). Sydney, Australia: Surrey Beatty and Sons.

Schwaner, T. D. and Sarre, S. D. (1988). Body size of tiger snakes in Southern Australia, with particular reference to *Notechis ater serventyi* (Elapidae) on Chappell Island. *Journal of Herpetology*, 22, 24-33.

Schwaner, T. D. and Sarre, S. D. (1990). Body size and sexual dimorphism in mainland and island tiger snakes. *Journal of Herpetology*, 24, 320-322.

Shine, R. (1977a). Habitats, diets and sympatry in snakes: a study from Australia. *Canadian Journal of Zoology*, 55, 1118-1128.

Shine, R. (1977b). Reproduction in Australian elapid snakes II. Female reproductive cycles. *Australian Journal of Zoology*, 25, 655-666.

Shine, R. (1987). Ecological comparisons of island and mainland populations of Australian tigersnakes (*Notechis*: Elapidae). *Herpetologica*, 43, 233-240.

Stearns, S. C. (1989). The evolutionary significance of phenotypic plasticity. *BioScience*, 39, 436-445.

Steiner, U. K. and Van Buskirk, J. (2008). Environmental stress and the costs of whole-organism phenotypic plasticity in tadpoles. *Journal of Evolutionary Biology*, 21, 97-103.

Steinger, T., Roy, B. A. and Stanton, M. L. (2003). Evolution in stressful environments ii: adaptive value and costs of plasticity in response to low light in *Sinapis arvensis. Journal of Evolutionary Biology*, 16, 313–323.

Swain, R. and Jones, S. M. (2000). Maternal effects associated with gestation conditions in a viviparous lizard, *Niveoscincus metallicus. Herpetological Monographs,* 14, 432-440.

Via, S., Gomulkiewicz, R., De Jong, G., Scheiner, S. M., Schlichting, C. D. and Van Tienderen, P. H. (1995). Adaptive phenotypic plasticity: consensus and controversy. *Trends Ecology and Evolution* 10, 212-217.

Vincent, S. E. and Mori, A. (2008). Determinants of feeding performance in free-ranging pit-vipers (Viperidae: *Ovophis okinavensis*): key roles for head size and body temperature. *Biological Journal of The Linnean Society*, 93, 53-62.

Waddington, C. H. (1942). Canalization of development and the inheritance of acquired characters. *Nature*, 150, 563-565.

Waddington, C. H. (1953). Genetic assimilation of an acquired character. *Evolution*, 7, 118-126.

Waddington, C. H. (1961). Genetic assimilation. *Advances in Genetics*, 10, 257-293.

Webb, J. K. and Whiting, M. J. (2005). Why don't small snakes bask? Juvenile broad-headed snakes trade thermal benefits for safety. *Oikos*, 110, 515-522.

Weinig, C., Johnston, J., German, Z. M. and Demink, L. M. (2006). Local and global costs of adaptive plasticity to density in *Arabidopsis thaliana. American Naturalist*, 167, 826-836.

West-Eberhard, M. J. (2003). *Development plasticity and evolution.* Oxford, UK: Oxford University Press.

Wilson, S. and Swan, G. (2003). A complete guide to reptiles of Australia. Sydney, Australia: New Holland Publishers.

Yeh, P. J. and Price, T. D. (2004). Adaptive phenotypic plasticity and the successful colonization of a novel environment. *The American Naturalist* 164, 531-542.

Reviewed by Jean Clobert, Head of the Station d'Ecologie Expérimentale à Moulis, 09200 Moulis, France

Fabien Aubret contribution towards "Phenotypic Plasticity: Molecular Mechanisms, Evolutionary Significance and Impact on Speciation"

In: Phenotypic Plasticity   ISBN: 978-1-62618-404-6
Editors: J. Valentino and P. Harrelson   © 2013 Nova Science Publishers, Inc.

Chapter 2

PHENOTYPIC PLASTICITY IN FUNCTIONAL TRAITS OF WOODY SPECIES IN TROPICAL DRY FORESTS

R. K. Chaturvedi[*] and A. S. Raghubanshi

Institute of Environment and Sustainable Development,
Banaras Hindu University, Varanasi, India

ABSTRACT

We analysed the phenotypic plasticity in nineteen functional traits (FTs) (seven morphological and twelve physiological) in tree and shrub species across the five study sites located in a tropical dry forest (TDF), showing variable soil moisture content (SMC). The aim was to observe the range of FTs in tree and shrub species across the study sites. Further, the response of FTs to SMC across species and sites was analysed. We also studied the relationships among FTs and between FTs and soil properties across sites. Results showed that the plasticity in FTs significantly varied across the study sites. The plasticity in FTs also differed significantly across species. All FTs under study affect relative growth rate (RGR) of the tree and shrub species directly or indirectly. However, the strength of effect is determined by environmental parameters and in case of TDF soil water availability is the important parameter. Plasticity in FTs due to changes in environmental parameters

[*] Corresponding Author: Name: Ravi K. Chaturvedi. Email: ravikantchaturvedi10@gmail.com; Phone: +91 9451584829; Fax: 0542 2368174.

explained the variations in RGR. Step-wise multiple regression indicates that more than 80% variability in RGR can be explained by canopy cover (CC), leaf area index (LAI), specific leaf area (SLA) and leaf intrinsic water use efficiency (WUEi) alone. First three variables represent quantity of photosynthetic surface and last represent water use economy of a species. All these are also significantly modulated by soil moisture availability. Important point to note here is that photosynthetic rate (A_{max}) is not an important parameter to determine RGR in TDF where water economy and extended period of leaflessness are critical.

Keywords: Phenotypic plasticity, Tropical dry forest, Functional traits, Soil water content, Specific leaf area, Photosynthetic rate, Stomatal conductance, Step-wise multiple regression

INTRODUCTION

The major goal of comparative plant ecology is to know how functional traits vary among species at different habitats and to what extent this variation has adaptive value (Poorter et al. 2008). Plant species vary widely in morphological and physiological traits, despite their shared key functional purpose of photosynthetic carbon assimilation and transpiration (Aerts & Chapin 2000; Reich & Oleksyn 2004; Wright et al. 2004, 2005). This capacity of a given plant species to express different phenotypes in different environments is known as phenotypic plasticity of the plant species (Sultan 2000). The plasticity of any given trait, which has a genetic basis and which may or may not be adaptive, can intensify or attenuate evolved responses, and can itself evolve in response to selection depending on the scale of spatial or temporal heterogeneity (Callahan 2005). Since selection can act on both variation for traits and variation for the plasticity of traits, the evolution of plastic traits can only be understood by addressing the complex interplay between the plastic responses of individuals and the evolved responses of populations (Callahan 2005).

Plasticity is an important aspect of evolution, development and function of plants in their environments and therefore, recently it has been widely recognized as a significant mode of plant functional trait diversity (Mathieu *et al.* 2009). Focusing on plant adjustments to nutrient stress, plasticity in terrestrial plants consists of low nutrient requirements, slow growth rates, high nutrient-use efficiency and uptake capacity and several modifications in root system morphology and biomass in order to improve nutrient uptake from the soil (Mony et al. 2007). An important type of potentially adaptive plasticity involves differences in the morphological, anatomical and physiological characteristics of leaves along environmental gradients such as light and or water availability.

Variation in leaf traits can be found across species (guilds, Givnish 1987), among populations of the same species (ecotypic differentiation and/or across-individual plasticity, Clausen et al. 1940), and even between leaves produced by a single plant (within-individual plasticity, Winn & Evans 1991; Winn 1996a,b). Furthermore, similar modifications of leaf structure and form in response to the environment appear at each of these levels (across or within species, populations or individuals). These traits may reasonably be expected to influence many of the generalized aspects of leaf function (e.g. stomatal density and rates of water and/or gas exchange, chlorophyll concentration and photosynthetic efficiency, Lewis 1972). In general, adaptive plasticity hypotheses predict that individuals capable of exhibiting such differences in leaf architecture in response to heterogeneous light environments should have higher fitness relative to less plastic individuals, however there are few tests of this prediction in the literature (but see Sultan & Bazzaz 1993; Winn 1999).

Extensive studies on plasticity of traits have generally been performed by comparing the responses of different species to variation in a single resource (Bradshaw 1965; Huber 1996; Poorter 1999; Robinson & Rorison 1988; Ryser & Eek 2000; Valladares et al. 2000). These studies have documented a large inter-specific variability in plasticity of morphological and structural traits. How consistent these differences may be in relation to different resources is still unclear, since plasticity levels and correlations between traits have been shown to change among environments (Schlichting 1989).

According to Wright et al. (2007), evolutionary ecologists are more concerned with the interspecific correlations among ecologically important plant traits because they may reflect two distinct phenomena. First, they may indicate physical, physiological or developmental 'constraints' that limit the independent variation and evolution of the representative traits for a particular environment. Second, the correlations may be the adaptive outcome of natural selection favouring particular combinations of traits over others, in which case the set of traits are often described as forming an ecological 'strategy' dimension (Westoby et al. 2002; Wright et al. 2007). It is important to distinctly perceive these explanations and understand the basis for trait-based strategy dimensions because it gives us insight into life-history trade-offs that operate between and within environments, and thus also into phenomena such as species coexistence, niche differentiation and the broad shifts in plant traits that occur along geographic gradients (Wright et al. 2007).

On the basis of literature survey, seven morphological (*viz.*, height, H; bark thickness, BT; wood specific gravity, WSG; leaf area, LA; canopy cover, CC; canopy depth, CD; leaf area index, LAI) and twelve physiological (*viz.*, leaf relative water content, RWC; leaf dry matter content, LDMC; specific leaf area, SLA; leaf carbon concentration, LCC; leaf nitrogen concentration, LNC; leaf phosphorus concentration, LPC; chlorophyll concentration, Chl; stomatal conductance, Gs_{max}; photosynthetic rate, A_{max}; intrinsic water use efficiency, WUEi; biomass increment, Bio Incr; relative growth rate, RGR) traits were selected which are important for the plant species of the dry deciduous forest. We observed their range in tree and shrub species across the study sites. Further, the response of functional traits to soil moisture content across species and sites was analysed. We also studied the relationships among functional traits and between functional traits and soil properties across sites.

METHODS

Study Sites

The study was conducted in five sites, Hathinala West (24° 18′ 07″ N and 83° 05′ 57″ E, 291 m.a.s.l.), Gaighat (24° 24′ 13″ N and 83° 12′ 01″ E, 245 m.a.s.l.), Harnakachar (24° 18′ 33″ N and 83° 23′ 05″ E, 323 m.a.s.l.), Ranitali (24° 18′ 11″ N and 83° 04′ 22″ E, 287 m.a.s.l.) and Kotwa (25° 00′ 17″ N and 82° 37′ 38″ E, 196 m.a.s.l.). Hathinala, Gaighat, Harnakachar and Ranitali sites are situated in Sonebhadra district and Kotwa in Mirzapur district of Uttar Pradesh (Figure 1). They occupy land area of 2555, 394, 1507, 2118 and 199 hectares, respectively. The sites were selected to represent a range in soil water availability. The forests on all the five sites are old-growth forests but have experienced disturbance in the form of grazing and illegal felling. These forests have been traditionally managed through selection felling, i.e., harvesting of individuals above a certain diameter which varies from species to species and leaving a few mother trees for regeneration (Singh & Singh 2011). Low intensity ground fire occurs every 2–3 years (Singh & Singh 2011). The area experiences tropical monsoon climate with three seasons in a year, *viz.* summer (April to mid June), rainy (mid June to September) and winter (November to February). The months of March and October constitute transition periods, respectively between winter and summer, and between rainy and winter seasons. The maximum monthly temperature varies from 20°C in January to 46°C in June, and the mean minimum monthly temperature reaches 12°C in January and 31°C in May. According to the data collected from the meteorological stations of the state forest department for 1980-2010, the mean annual rainfall ranges from 1196 mm (Hathinala) to 865 mm (Kotwa site) (Chaturvedi et al. 2011). About 85% of the annual rainfall occurs during the rainy season from the south-west monsoon, and the remaining from the few showers in December and in May-June. There is an extended dry period of about 9 months (October-mid June) in the annual cycle (Chaturvedi et al. 2011). The monthly rainfall varies from 6 mm in April to 334 mm in August.

Study Design and Field Methods

At each site, we randomly selected three 4 ha (200 × 200 m) plots and for the estimation of importance value index (IVI) of tree (> 30 cm dbh) and shrub species, we randomly placed three 1000 m^2 (50 × 20 m) sub-plots in each selected plot. IVI was calculated as the average of the values for relative basal area, relative density and relative frequency (Curtis & McIntosh 1951). Composite surface (0 – 30 cm) soil samples were also collected, but only once, from each sub-plot for physico-chemical analysis. These samples were analysed for texture (Sheldrick & Wang 1993), organic carbon (Walkley & Black 1934), total nitrogen (Bremner & Mulvaney 1982) and total phosphorus (Olsen & Sommers 1982) contents. Five mature individuals of each woody species were randomly selected from the three 4 ha plot at each site and marked with permanent marker. Thus, at Hathinala 205 (41 × 5), at Gaighat 160 (32 × 5), at Harnakachar 170 (34 × 5), at Ranitali 145 (29 × 5) and at Kotwa 70 (14 × 5) individuals were marked. Near each marked individual, at a distance of one metre from the stem, soil moisture content was estimated in three directions with Delta-T theta probe moisture meter. Their girth was recorded and other morphological traits were estimated as described in Chaturvedi & Raghubanshi (2011). For leaf traits, healthy and mature leaves were selected from each individual and analysed according to protocol described in Chaturvedi & Raghubanshi (2011). Biomass increment and relative growth rate was studied by taking initial girth in July 2005 and final girth values in June 2006 and June 2007. The above mentioned morphological and leaf traits were studied during the months of September and October in the year 2005.

Plasticity of plant traits across sites were calculated by using the formula according to Callahan (2005) as:

$$Trait\ plasticity = \left[1 - \frac{lowest\ value\ of\ plant\ trait\ across\ sites}{highest\ value\ of\ plant\ trait\ across\ sites}\right] \times 100$$

Data of SMC and twenty plant traits were analyzed by MANOVA. Two-tailed Pearson correlation coefficients among SMC and FTs were calculated. All the statistical analyses were done using SPSS (ver. 16) package. Effect of soil properties on seven morphological and twelve physiological traits of the plant species were studied with the help of regression plots prepared by using SigmaPlot (ver.11).

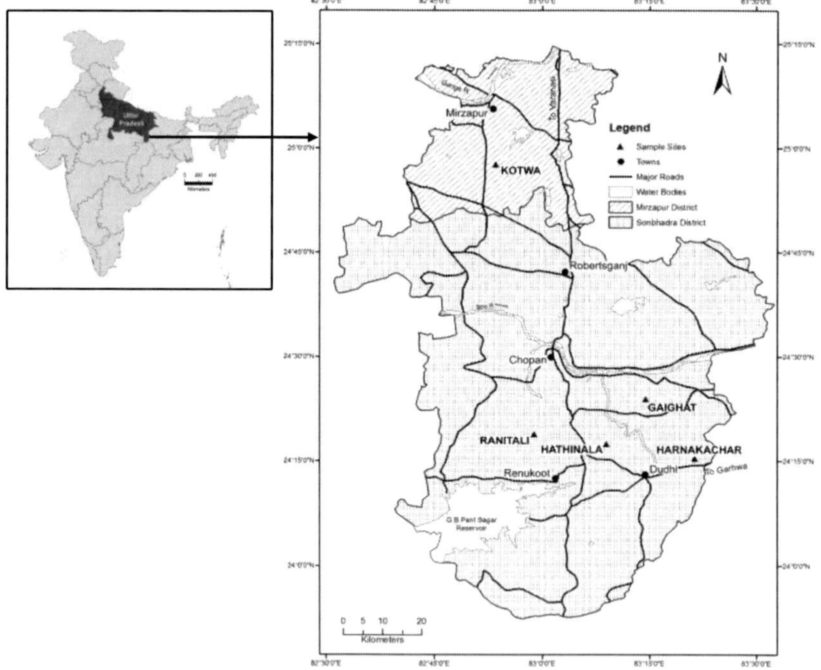

Figure 1. Map showing the location of study areas.

RESULTS

Site Characteristics

Soil analysis indicated greatest annual average SMC, clay content, bulk density, organic C, total N and total P at the Hathinala site and the lowest at the Kotwa site (Table 1). Significant effect of site was indicated for SMC, clay, silt, sand, bulk density, organic C, total N and total P. Mean values of organic C, total N and total P averaged across the five sites were 21.8, 1.71 and 0.43 t ha^{-1}. IVI of tree and shrub species of the study sites are given in Table 2.

Table 1. Physiochemical properties of the soil, species richness and density of tree saplings at the five study sites located in the tropical dry forest

Parameters	Hathinala	Gaighat	Harnakachar	Ranitali	Kotwa
Annual average	13.0^a	11.1^{a,b}	10.2^b	7.78^c	6.41^c
SMC (%)	(±0.68)	(±0.71)	(±0.39)	(±0.62)	(±0.46)
Clay	10.6^a	7.08^b	4.83^c	3.00^d	2.12^d
(%)	(±0.68)	(±0.50)	(±0.25)	(±0.31)	(±0.21)
Silt	22.9^a	32.1^b	26.2^c	26.8^c	27.2^c
(%)	(±2.30)	(±0.81)	(±0.36)	(±0.70)	(±0.50)
Sand	66.5^a	60.8^b	69.0^c	70.2^c	70.7^c
(%)	(±0.33)	(±0.57)	(±0.35)	(±0.49)	(±0.35)
Organic C	1.89^a	1.61^{a,b}	1.56^{a,b}	1.39^b	1.20^b
(%)	(±0.07)	(±0.06)	(±0.09)	(±0.18)	(±0.10)
Total N	0.14^a	0.13^{a,b}	0.13^{a,b}	0.12^{a,b}	0.11^b
(%)	(±0.01)	(±0.01)	(±0.01)	(±0.01)	(±0.00)
Total P	0.04^a	0.03^{a,b}	0.03^{a,b}	0.02^{a,b}	0.01^b
(%)	(±0.01)	(±0.00)	(±0.00)	(±0.00)	(±0.00)

(Mean ± 1S.E.) Different superscript letters indicate significant difference ($P < 0.05$) from each other.

Functional Traits

Morphological Traits

In trees, maximum GT (cm), HT (17 m), BT (1.8 cm) and LAI (14) were observed in *S. robusta*, greatest WSG (0.80 g cm^{-3}), CC (28 m^2) and CD (7.8 m) in *H. binata* and largest LA (665 cm^2) in *S. urens* (Table 3). In shrubs, maximum GT (24 cm), HT (4 m), BT (0.5 cm), LA (31 cm^2), CC (13 m^2) and

CD (3 m) were detected in *L. camara*, highest WSG (0.7 g cm^{-3}) in *C. spinarum* and greatest LAI (3.9) in *W. fruticosa* (Table 4).

Table 2. Importance value index (IVI) ± 1 SE of tree and shrub species in the study sites

S.No.	Species	Hathinala (IVI ± SE)	Gaighat (IVI ± SE)	Harnaka-char (IVI ± SE)	Ranitali (IVI ± SE)	Kotwa (IVI ± SE)
	Tree					
	Acacia auriculiformis	-	4.4±1.3	-	-	-
	Acacia catechu	6.7±2.4	1.0±0.5	-	21.0±3.4	18.1±5.3
	Adina cordifolia	3.1±3.1	-	-	-	-
	Albizia odoratissima	0.3±0.3	-	-	-	-
	Anogeissus latifolia	4.1±1.2	7.0±1.1	8.8±1.4	10.2±1.8	-
	Azadirachta indica	-	3.2±1.1	-	-	-
	Bauhinia racemosa	1.4±0.8	-	-	-	-
	Boswellia serrata	4.2±2.4	2.8±1.5	1.0±1.0	3.3±3.3	-
	Briedelia retusa	2.3±1.7	-	-	-	-
	Buchanania lanzan	8.9±2.0	14.1±0.5	9.5±5.3	4.9±4.9	-
	Cassia fistula	0.4±0.4	-	-	-	-
	Cassia siamea	-	2.8±2.1	-	-	-
	Diospyros melanoxylon	6.3±0.9	8.3±0.4	8.5±0.4	9.8±1.0	2.6±1.5
	Elaeodendron glaucum	0.3±0.3	3.6±1.0	0.8±0.8	-	-
	Emblica officinalis	0.8±0.8	2.3±1.4	2.4±0.2	1.1±1.1	3.4±2.6
	Ficus racemosa	-	-	-	-	18.0±3.4
	Flacourtia indica	0.3±0.3	1.0±0.5	0.5±0.5	-	-
	Gardenia latifolia	2.8±1.6	-	1.1±0.6	-	-
	Gardenia turgida	0.3±0.3	-	-	-	-
	Grewia serrulata	0.3±0.3	-	0.5±0.5	-	-
	Hardwickia binata	7.1±4.7	0.7±0.7	3.5±2.5	8.3±4.7	-
	Hollarrhena andidysenterica	-	-	1.1±0.6	-	4.8±2.2
	Holoptelia integrifolia	0.1±0.1	0.6±0.6	1.1±0.6	-	-
	Lagerstroemia parviflora	9.0±2.7	7.7±4.7	8.4±4.8	13.0±2.5	2.3±1.3
	Lannea coromandelica	1.5±1.0	2.4±1.6	1.7±1.7	1.4±0.7	21.2±4.7
	Madhuca longifolia	-	-	5.9±3.5	-	-
	Miliusa tomentosa	2.5±1.3	-	1.4±0.7	-	-
	Mitragyna parvifolia	1.8±0.9	-	-	-	-
	Nyctanthes arbortristis	-	-	-	-	12.5±5.1
	Pterocarpus marsupium	1.4±0.8	-	0.8±0.8	-	-
	Schleichera oleosa	0.5±0.5	-	-	-	-
	Semecarpus anacardium	0.3±0.3	-	-	-	-
	Shorea robusta	14.0±3.0	19.5±1.1	13.7±2.4	4.1±2.8	-
	Soymida febrifuga	7.5±3.0	-	10.0±0.8	8.0±4.7	-
	Sterculia urens	-	-	-	-	5.4±3.1
	Terminalia chebula	0.3±0.3	-	-	-	-
	Terminalia tomentosa	9.6±2.8	12.5±2.2	10.0±1.2	9.6±1.3	-

S.No.	Species	Hathinala (IVI ± SE)	Gaighat (IVI ± SE)	Harnaka-char (IVI ± SE)	Ranitali (IVI ± SE)	Kotwa (IVI ± SE)
	Zizyphus glaberrima	0.3±0.3	-	-	-	8.7±1.5
	Zizyphus nummularia	0.5±0.5	1.1±0.6	0.6±0.6	-	-
	Shrub					
	Carissa spinarum	-	0.5±0.5	1.9±1.2	-	-
	Lantana camara	-	0.5±0.5	-	-	-
	Woodfordia fruticosa	-	3.4±2.3	2.0±0.4	0.9±0.9	-
	Zizyphus oenoplea	-	-	3.0±1.4	2.5±1.2	3.0±1.7
	Grewia hirsuta	1.1±0.6	0.6±0.3	1.8±0.2	1.9±0.5	-

Table 3. Range of morphological traits of tree species (n = 39) across sites

Trait	Min	Max	Mean (±S.E.)	Plasticity (%)
GT (cm)	36.5 (*Z. nummularia*, HK)	104 (*S. robusta*, HN)	64.7 (±2.1)	64.7
HT (m)	4.10 (*Z. nummularia*, HK)	16.7 (*S. robusta*, HN)	7.80 (±0.4)	75.4
BT (cm)	0.60 (*A. latifolia*, HN)	1.80 (*S. robusta*, GG)	1.20 (±0.1)	66.7
WSG (g cm^{-3})	0.40 (*S. urens*, KT)	0.80 (*H. binata*, RT)	0.60 (±01)	50.0
LA (cm^2)	5.00 (*Z. glaberrima*, KT)	665 (*S. urens*, HN)	155 (±28.4)	99.2
CC (m^2)	9.40 (*F. indica*, GG)	28.4 (*H. binata*, RT)	16.0 (±0.5)	66.9
CD (m)	1.80 (*F. indica*, HN)	7.80 (*H. binata*, RT)	4.20 (±0.2)	76.9
LAI (m^2 m^{-2})	2.80 (*G. turgida*, HN)	14.0 (*S. robusta*, HN)	7.60 (±0.4)	80.0

GT, girth; HT, height; BT, bark thickness; WSG, wood specific gravity; LA, leaf area; CC, canopy cover; CD, canopy depth; LAI, leaf area index; HN, Hathinala; GG, Gaighat; HK, Harnakachar; RT, Ranitali; KT, Kotwa.

Physiological Traits

In trees, maximum SLA (160 cm^2 g^{-1}), LNC (2.5%), LPC (0.4%), Chl (2.0 mg g^{-1}), Gs_{max} (0.34 mol m^{-2} s^{-1}) and A_{max} (15 µmol m^{-2} s^{-1}) were observed in *T. tomentosa*, highest RWC (99%) in *F. indica*, greatest LDMC (38%) in *H. binata*, maximum LCC (46%) in *F. racemosa*, highest WUEi (62 µmol mol^{-1}) and RGR (0.09 cm^2 cm^{-2} yr^{-1}) in *Z. glaberrima* and greatest Bio Incr (2.1 kg mo^{-1}) in *A. cordifolia* (Table 5). In shrubs, maximum RWC (95%), SLA (159 cm^2 g^{-1}), LNC (2.2%), Chl (1.9 mg g^{-1}), Gs_{max} (0.3 mol m^{-2} s^{-1}), A_{max} (14 µmol m^{-2} s^{-1}) and Bio Incr (0.11 kg mo^{-1}) were analyzed in *L. camara*, greatest LDMC (37%) and LCC (47%) in *G. hirsuta*, highest LPC (0.3%) and WUEi (65 µmol mol^{-1}) in *Z. oenoplea* and maximum RGR (0.16 cm^2 cm^{-2} yr^{-1}) in *W. fruticosa* (Table 6).

Table 4. Range of morphological traits of shrub species (n = 5) across sites

Trait	Min	Max	Mean (±S.E.)	Plasticity (%)
GT (cm)	7.50 (*G. hirsuta*, RT)	24.4 (*L. camara*, HN)	15.6 (±2.5)	69.3
HT (m)	0.80 (*G. hirsuta*, RT)	3.70 (*L. camara*, HN)	1.50 (±0.3)	78.4
BT (cm)	0.30 (*Z. oenoplea*, KT)	0.50 (*L. camara*, HK)	0.41 (±0.02)	40.0
WSG (g cm^{-3})	0.60 (*L. camara*, HN)	0.70 (*C. spinarum*, HK)	0.61 (±0.02)	14.3
LA (cm^2)	0.80 (*Z. oenoplea*, KT)	31.4 (*L. camara*, HN)	6.72 (±5.6)	97.5
CC (m^2)	0.70 (*G. hirsuta*, GG)	12.9 (*L. camara*, HN)	3.20 (±1.3)	94.6
CD (m)	0.60 (*G. hirsuta*, HK)	3.10 (*L. camara*, HN)	1.20 (±0.2)	80.6
LAI (m^2 m^{-2})	1.20 (*G. hirsuta*, HN)	3.90 (*W. fruticosa*, HK)	2.81 (±0.4)	69.2

GT, girth; HT, height; BT, bark thickness; WSG, wood specific gravity; LA, leaf area; CC, canopy cover; CD, canopy depth; LAI, leaf area index; HN, Hathinala; GG, Gaighat; HK, Harnakachar; RT, Ranitali; KT, Kotwa.

PLASTICITY OF FTS ACROSS SITES

In the case of morphological traits, highest trait plasticity was observed in leaf area (LA) (99.2% in trees and 97.5% in shrubs) and lowest in wood specific gravity (WSG) (50.0% in trees and 14.3% in shrubs) (Table 3). Lowest LA was detected in *Z. glaberrima* (5.0 cm^2) at Kotwa and highest in *S. urens* (665 cm^2) at Hathinala, whereas, WSG was minimum in *S. urens* (0.4 g cm^{-3}) at Kotwa and maximum in *H. binata* (0.8 g cm^{-3}) at Ranitali (Table 3). Lowest LA in shrubs was observed in *Z. oenoplea* (0.8 cm^2) at Kotwa and highest in *L. camara* (31.4 cm^2) at Hathinala (Table 4). WSG of shrub species was minimum in *L. camara* (0.6 g cm^{-3}) at Hathinala and maximum in *C. spinarum* (0.7 g cm^{-3}) at Harnakachar (Table 4).

Table 5. Range of physiolological traits of tree species (n = 39) across sites

Trait	Min	Max	Mean (±S.E.)	Plasticity (%)
RWC (%)	68.6 (*L. coromandelica*, KT)	98.9 (*F. indica*, HN)	93.1 (±1.0)	30.6
LDMC (%)	33.0 (*G. turgida*, HN)	37.7 (*H. binnata*, RT)	35.0 (±0.2)	12.5
SLA ($cm^2 g^{-1}$)	61.3 (*F. racemosa*, KT)	159.6 (*T. tomentosa*, HN)	120 (±4.6)	61.6
LCC (%)	42.4 (*L. coromandelica*, HN)	46.4 (*F. racemosa*, KT)	44.6 (±0.1)	8.62
LNC (%)	1.30 (*Z. nummularia*, HN)	2.50 (*T. tomentosa*, RT)	1.83 (±0.03)	48.0
LPC (%)	0.10 (*B. lanzan*, RT)	0.40 (*T. tomentosa*, GG)	0.22 (±01)	75.0
Chl ($mg g^{-1}$)	0.60 (*B. lanzan*, RT)	2.00 (*T. tomentosa*, HN)	1.21 (±04)	70.0
Gs_{max} ($mol\ m^{-2}\ s^{-1}$)	0.20 (*D. melanoxylon*, KT)	0.34 (*T. tomentosa*, HN)	0.26 (±0.01)	41.2
A_{max} ($\mu\ mol\ m^{-2}\ s^{-1}$)	4.60 (*G. turgida*, HN)	15.3 (*T. tomentosa*, RT)	11.4 (±0.3)	69.9
WUEi ($\mu\ mol\ mol^{-1}$)	22.9 (*A. cordiafolia*, HN)	62.1 (*Z. glaberrima*, KT)	46.1 (±1.1)	63.1
Bio Incr ($kg\ mo^{-1}$)	0.20 (*S. robusta*, HN)	2.10 (*A. cordiafolia*, HN)	0.76 (±0.05)	90.5
RGR ($cm^2\ cm^{-2}\ yr^{-1}$)	0.005 (*S. robusta*, HN)	0.09 (*Z. glaberrima*, KT)	0.04 (±0.003)	94.4

RWC, relative water content; LDMC, leaf dry matter content; SLA, specific leaf area; LCC, leaf carbon concentration; LNC, leaf nitrogen concentration; LPC, leaf phosphorus concentration; Chl, chlorophyll concentration; Gs_{max}, stomatal conductance; A_{max}, photosynthetic rate; WUEi, intrinsic water use efficiency; Bio Incr, biomass increment; RGR, relative growth rate; HN, Hathinala; GG, Gaighat; HK, Harnakachar; RT, Ranitali; KT, Kotwa.

Trait plasticity of physiological traits was greatest for relative growth rate (RGR) (94.4%) in tree species and for biomass increment (Bio Incr) (90.0%) in shrubs (Table 5 & 6). Lowest plasticity in trees was observed for leaf carbon concentration (LCC) (8.62%) and in shrubs for leaf dry matter content (LDMC) (6.54%) (Table 5 & 6). Maximum RGR in tree species was estimated in *Z. glaberrima* (0.09 cm^2 cm^{-2} yr^{-1}) at Kotwa and minimum in *S. robusta* (0.005 cm^2 cm^{-2} yr^{-1}) at Hathinala. Greatest Bio Incr in shrubs was observed in *L. camara* (0.11 $kg\ mo^{-1}$) at Hathinala and lowest in *G. hirsuta* (0.01 $kg\ mo^{-1}$) also at Hathinala (Table 5 & 6). Greatest LCC in tree species was observed in *F. racemosa* (46.4%) at Kotwa and lowest in *L. coromandelica* (42.4%) at

Hathinala. In shrubs, highest LDMC was found in *G. hirsuta* (36.7%) at Ranitali and lowest in *Z. oenoplea* (34.3%) at Hathinala (Table 5 & 6).

Table 6. Range of physiolological traits of shrub species (n = 5) across sites

Trait	Min	Max	Mean (±S.E.)	Plasticity (%)
RWC (%)	81.8 (*W. fruticosa*, RT)	95.4 (*L. camara*, HN)	90.1 (±1.5)	14.3
LDMC (%)	34.3 (*Z. oenoplea*, HN)	36.7 (*G. hirsuta*, RT)	35.7 (±0.2)	6.54
SLA ($cm^2\ g^{-1}$)	72.1 (*G. hirsuta*, RT)	159 (*L. camara*, HN)	116 (±11.3)	54.6
LCC (%)	43.1 (*L. camara*, HN)	46.8 (*G. hirsuta*, RT)	45.1 (±0.4)	7.91
LNC (%)	1.30 (*G. hirsuta*, HN)	2.20 (*L. camara*, RT)	1.51 (±0.1)	40.9
LPC (%)	0.10 (*G. hirsuta*, RT)	0.30 (*Z. oenoplea*, HN)	0.20 (±0.02)	66.7
Chl ($mg\ g^{-1}$)	0.50 (*G. hirsuta*, RT)	1.90 (*L. camara*, HN)	0.92 (±0.1)	73.7
Gs_{max} ($mol\ m^{-2}\ s^{-1}$)	0.10 (*G. hirsuta*, RT)	0.30 (*L. camara*, HN)	0.18 (±0.02)	66.7
A_{max} ($\mu\ mol\ m^{-2}\ s^{-1}$)	6.00 (*G. hirsuta*, HN)	13.7 (*L. camara*, HN)	9.89 (±1.2)	56.2
WUEi ($\mu\ mol\ mol^{-1}$)	45.3 (*Z. oenoplea*, HN)	65.2 (*Z. oenoplea*, RT)	55.8 (±1.0)	30.5
Bio Incr ($kg\ mo^{-1}$)	0.01 (*G. hirsuta*, HN)	0.11 (*L. camara*, HN)	0.06 (±0.01)	90.9
RGR ($cm^2\ cm^{-2}\ yr^{-1}$)	0.05 (*L. camara*, HN)	0.16 (*W. fruticosa*, RT)	0.10 (±0.01)	68.8

RWC, relative water content; LDMC, leaf dry matter content; SLA, specific leaf area; LCC, leaf carbon concentration; LNC, leaf nitrogen concentration; LPC, leaf phosphorus concentration; Chl, chlorophyll concentration; Gs_{max}, stomatal conductance; A_{max}, photosynthetic rate; WUEi, intrinsic water use efficiency; Bio Incr, biomass increment; RGR, relative growth rate; HN, Hathinala; GG, Gaighat; HK, Harnakachar; RT, Ranitali; KT, Kotwa.

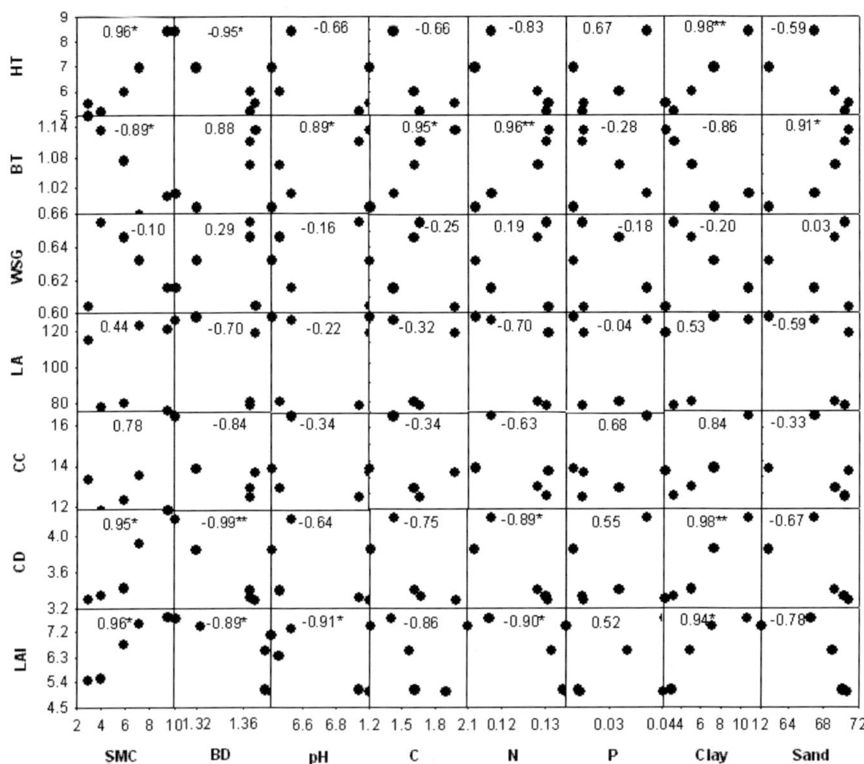

Figure 2. Relationships of morphological traits and soil properties of the five study sites. HT, height; BT, bark thickness; WSG, wood specific gravity; LA, leaf area; CC, canopy cover; CD, canopy depth; LAI, leaf area index; SMC, soil moisture content; BD, bulk density; C, organic carbon content; N, total nitrogen content; P, total phosphorus content; *$P < 0.05$; **$P < 0.01$. Numbers indicate R values.

with the plant HT (Figure 2). Strongest relationship of bark thickness (BT) was with sand ($R = 0.91$, $P < 0.05$). This trait was positively related with BD, pH and sand content but negatively related with SMC, C, N, P and clay (Figure 2). WSG showed negative relationship with SMC, pH, P and clay, whereas, positive relationship with BD. Positive relationship of LA was detected with SMC, P and clay content but negative relationship with BD, pH, C, N and sand (Figure 2). Plant canopy cover (CC) was positively related with SMC, C, N, P and clay but showed negative relationship with BD, pH and sand. Strongest relationship of canopy depth (CD) was observed with BD ($R = -0.99$, $P < 0.01$) which was significantly negative (Figure 2). The other soil properties showing negative relationship with CD were pH and sand but SMC, C, N, P and clay showed positive relationship with CD. Leaf area index (LAI) showed strongest relationship with SMC ($R = 0.96$, $P < 0.05$) as compared to other soil properties. SMC, C, N, P and clay were positively related with LAI, whereas, BD, pH and sand showed negative relationship with LAI (Figure 2).

Physiological Traits

The relationship of leaf relative water content (RWC) was strongest with SMC ($R = 0.95$, $P < 0.01$). It has been observed that among various soil properties, RWC was positively related with SMC, C, N, P and clay, however, it was negatively related with BD, pH and sand (Figure 3). LDMC also showed strongest relationship with SMC ($R = -0.96$, $P < 0.01$). Its relationship with SMC, N, P and clay were negative but the relationship with BD, pH, C and sand were positive (Figure 3). SLA was strongly related with SMC ($R = 0.99$, $P < 0.01$) and clay content ($R = 0.99$, $P < 0.01$). Its relationship with SMC and clay were positive, whereas, the relationship with BD, pH and sand were negative (Figure 3). LCC showed strongest relationship with clay ($R = -0.99$, $P < 0.01$) and it was detected that its relationship with SMC, C, N, P and clay were negative and that with BD, pH and sand were positive (Figure 3). The relationship of leaf nitrogen concentration (LNC) was strongest with pH ($R = 0.91$, $P < 0.05$). It was observed that the relationship of LNC with SMC, C, N, P and clay were negative and the relationship with BD, pH and sand were positive (Figure 3). Leaf phosphorus concentration (LPC) showed strongest relationship with clay ($R = 0.97$, $P < 0.01$). Its relationship with SMC, C, N, P and clay were positive and the relationship with BD, pH and sand were negative (Figure 3). The strength of relationship of leaf chlorophyll concentration (Chl) was strongest with clay ($R = 1.00$, $P < 0.01$). It was observed that the relationship with SMC, C, N, P and clay were positive, whereas, the relationship with BD, pH and sand were negative (Figure 3). Stomatal conductance (Gs_{max}) also showed strongest relationship with clay ($R = 0.97$, $P < 0.01$). Its relationship with SMC, C, N, P and clay were positive and the relationship with BD, pH and sand were negative (Figure 3). A_{max} showed negative relationship with SMC, P and clay and positive correlations with BD, pH, C, N and sand. The relationship of leaf intrinsic water use efficiency (WUEi) with clay was strongest ($R = -0.98$, $P < 0.01$) as compared to its relationships with other soil properties. It showed positive correlations with BD, pH, N and sand. Bio Incr was positively associated with BD, pH, C,

N and sand and negatively with SMC, P and clay. On the other hand RGR showed positive correlations with BD, pH and sand and negative with SMC, C, N, P and clay (Figure 3).

Table 7. Summary of MANOVA on soil moisture content (SMC) and functional traits (FTs) of woody species of the five study sites

Trait	Site ($F_{4,448}$)	Species ($F_{43,448}$)	Site × Species ($F_{64,448}$)
SMC	558***	32.9***	4.54***
GT	32.9***	102***	2.97***
HT	92.3***	76.3***	3.66***
BT	16.4***	52.4***	1.17^{ns}
WSG	22.7***	86.4***	1.41*
LA	89.3***	1701***	13.2***
CC	86.8***	107***	4.83***
CD	12.2***	37.4***	1.63**
LAI	44.9***	78.9***	4.59***
RWC	734***	415***	17.4***
LDMC	395***	40.5***	4.95***
SLA	661***	757***	5.76***
LCC	488***	66.6***	5.16***
LNC	47.3***	22.7***	1.88***
LPC	19.6***	4.92***	1.05^{ns}
Chl	311***	44.5***	3.79***
Gs_{max}	91.9***	17.6***	1.60**
A_{max}	19.0***	215***	17.7***
WUEi	190***	12.2***	1.64**
Bio Incr	29.3***	69.5***	3.79***
RGR	24.0***	17.1***	1.73**

GT, girth; HT, height; BT, bark thickness; WSG, wood specific gravity; LA, leaf area; CC, canopy cover; CD, canopy depth; LAI, leaf area index; RWC, relative water content; LDMC, leaf dry matter content; SLA, specific leaf area; LCC, leaf carbon concentration; LNC, leaf nitrogen concentration; LPC, leaf phosphorus concentration; Chl, chlorophyll concentration; Gs_{max}, stomatal conductance; A_{max}, photosynthetic rate; WUEi, intrinsic water use efficiency; Bio Incr, biomass increment; RGR, relative growth rate; $^{ns}P > 0.05$; $*P < 0.05$; $**P < 0.01$; $***P < 0.001$.

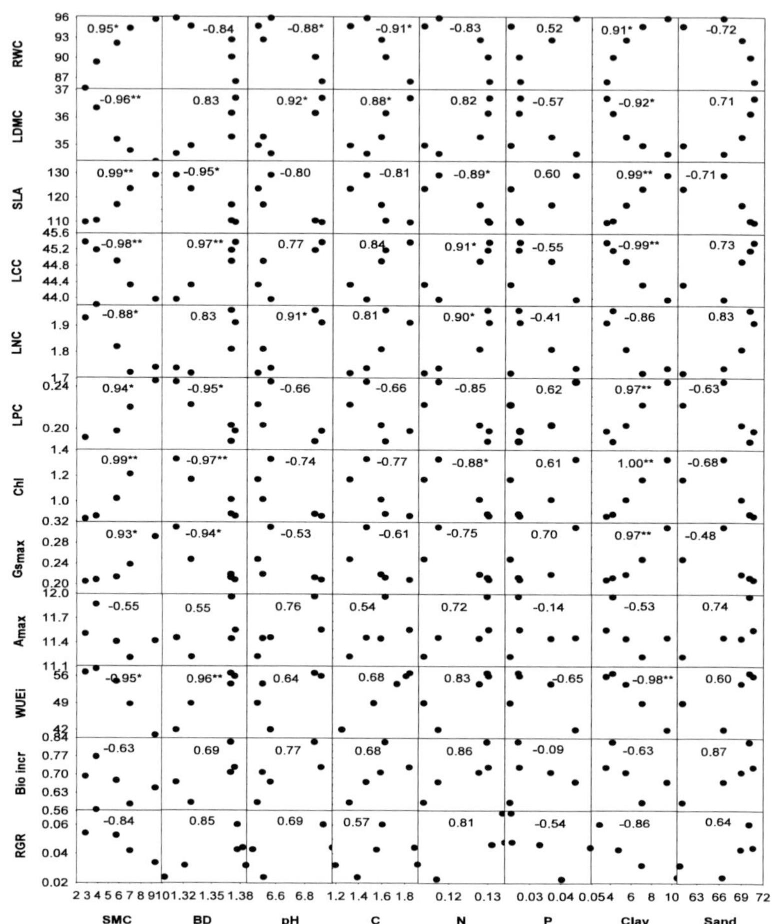

Figure 3. Relationships of physiological traits soil properties of the five study sites. RWC, relative water content; LDMC, leaf dry matter content; SLA, specific leaf area; LCC, leaf carbon concentration; LNC, leaf nitrogen concentration; LPC, leaf phosphorus concentration; Chl, chlorophyll concentration; Gs_{max}, stomatal conductance; A_{max}, photosynthetic rate; WUEi, intrinsic water use efficiency; Bio Incr, biomass increment; RGR, relative growth rate; SMC, soil moisture content; BD, bulk density; C, organic carbon content; N, total nitrogen content; P, total phosphorus content; $*P < 0.05$; $**P < 0.01$. Numbers indicate R values.

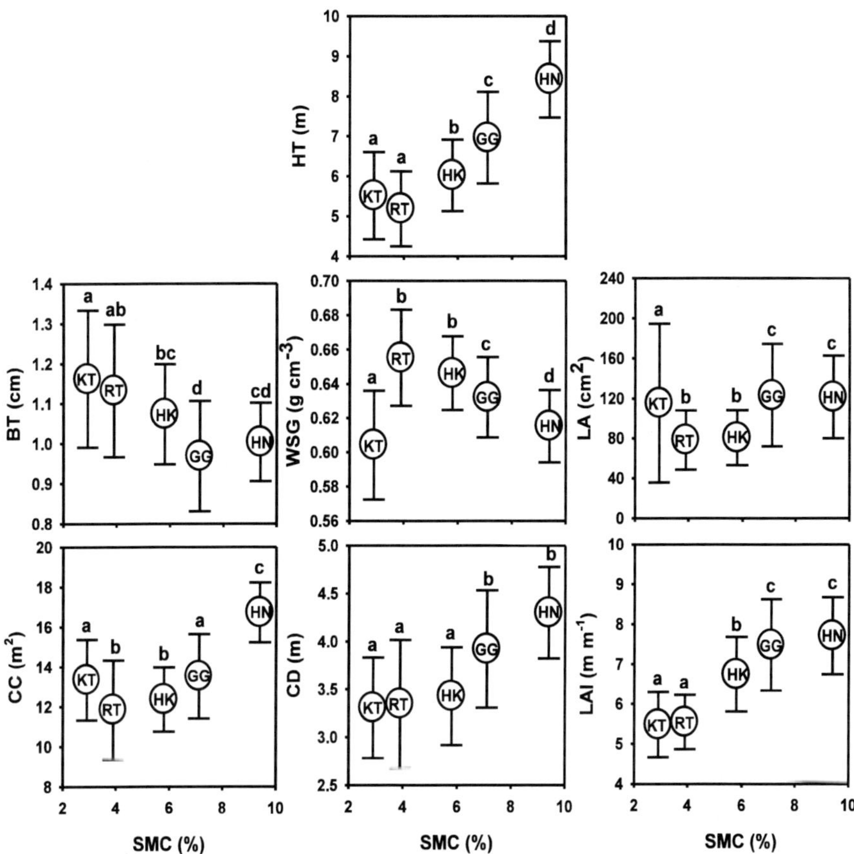

Figure 4. Relationship of morphological traits across study sites with SMC, soil moisture content. HT, height; BT, bark thickness; WSG, wood specific gravity; LA, leaf area; CC, canopy cover; CD, canopy depth; LAI, leaf area index; HN, Hathinala; GG, Gaighat; HK, Harnakachar; RT, Ranitali; KT, Kotwa.

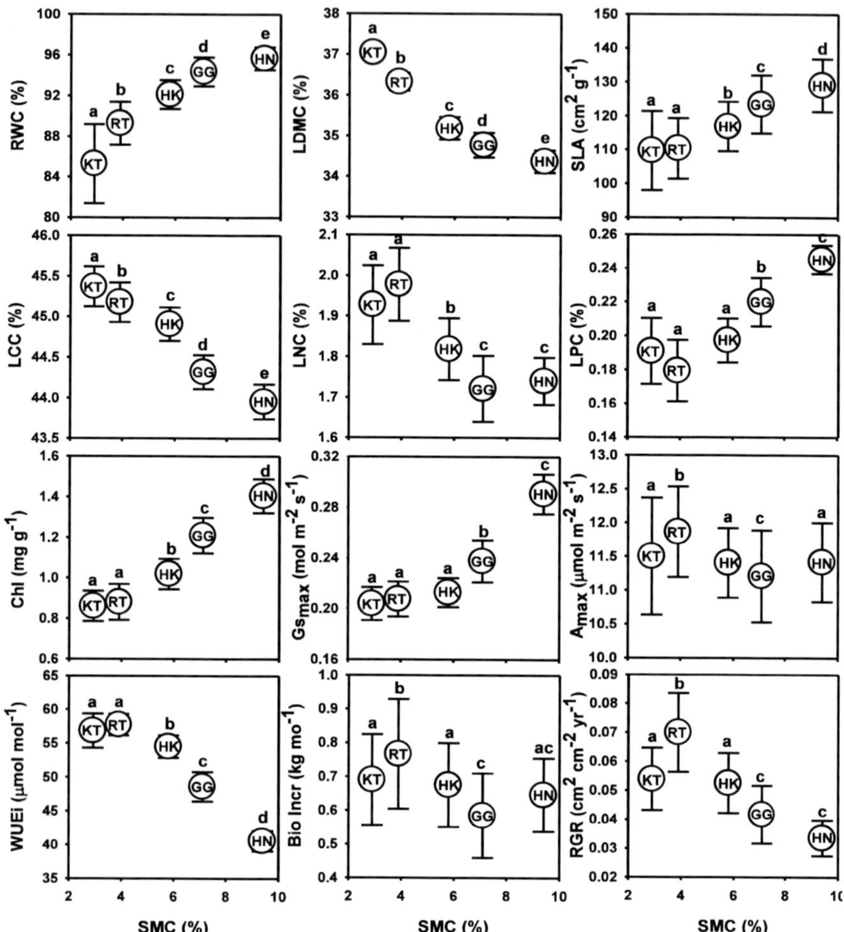

Figure 5. Relationship of physiological traits across study sites with SMC, soil moisture content. RWC, relative water content; LDMC, leaf dry matter content; SLA, specific leaf area; LCC, leaf carbon concentration; LNC, leaf nitrogen concentration; LPC, leaf phosphorus concentration; Chl, chlorophyll concentration; Gs_{max}, stomatal conductance; A_{max}, photosynthetic rate; WUEi, intrinsic water use efficiency; Bio Incr, biomass increment; RGR, relative growth rate; HN, Hathinala; GG, Gaighat; HK, Harnakachar; RT, Ranitali; KT, Kotwa.

FT VARIATIONS ACROSS SITE AND SPECIES

Functional traits averaged across species showed significant difference across sites (Table 7). Average SMC in the study sites were also significantly different ($F_{4,\ 448}$ = 558, P < 0.001) and the mean value ranged from 2.9 % at Kotwa to 9.4% at Hathinala (Figure 4). MANOVA of FTs across sites and species showed significant difference (Table 7). Two-way interaction of site and species were also significant for all FTs except for BT and LPC where interactions were statistically not significant.

Morphological Traits

Tukey post-hoc comparisons of the average values of morphological traits showed significant difference across sites at P < 0.05 (Figure 4). Among morphological traits, HT, CC, CD and LAI were greatest at Hathinala, which is the most moist site, whereas, BT was maximum at Kotwa which is the most dry site. In the remaining two traits, WSG was highest at Ranitali, also considered as a dry site and LA was greatest at Gaighat which is comparatively moist site among the five study sites.

Physiological Traits

The mean values of 12 physiological traits studied in the woody species at the five study sites, also showed significant difference across sites as explained by the Tukey post-hoc comparison (Figure 5). RWC, SLA, LPC, Chl and Gs_{max} were greatest at the most moist site, Hathinala, whereas, LDMC and LCC were maximum at the most dry site, Kotwa. The remaining five traits, i.e., LNC, A_{max}, WUEi, Bio Incr and RGR were highest at Ranitali which is also comparatively dry site next to Kotwa.

RELATIONSHIPS AMONG FTS AND BETWEEN FTS AND SMC

The study showed significant correlations among FTs and between FTs and SMC except for the relationship of GT with RWC and SLA, BT with WSG, RWC, LDMC, SLA, LCC, LPC, Chl, Gs_{max} and WUEi, WSG with LDMC, LCC, LPC, Chl, Gs_{max}, Bio Incr and RGR, LA with RWC, SLA, LCC, LNC, LPC, Chl and A_{max}, CC with LDMC, RWC with LNC, A_{max} and Bio Incr, LDMC with SLA, LNC and A_{max}, SLA with WUEi, Bio Incr and RGR,

LCC with LNC and Bio Incr, LNC with Chl and Gs_{max}, LPC, Chl and WUEi with Bio Incr and the relationship of SMC with LNC and Bio Incr (Table 8). Most of the relationships of FTs were positive except for the relationships of LDMC, LCC, WUEi and RGR, where, their relationships with other FTs were significantly negative, although, they were positively related with each other. Moreover, WUEi was positively correlated with WSG ($R = 0.22$, $P < 0.05$), LNC ($R = 0.25$, $P < 0.01$) and A_{max} ($R = 0.20$, $P < 0.05$).

DISCUSSION

Plasticity is a biological characteristic which has been widely recognized as an important feature of organism's development, function and evolution in their environments (Sultan 2000; Mony et al. 2007). This corresponds to the ability of an organism to adjust its performance by altering its physiology, morphology and life-history in response to variations in environmental conditions (Bradshaw et al. 1964; Schlichting 1986; Sultan 1995; Sultan 2000; Mony et al. 2007). The present study showed a wide range of plasticity in the woody species of the study sites. They also showed very high variation in their relationships with SMC and other soil physico-chemical properties.

Morphological plasticity plays an important role in resource acquisition of plants (Bradshaw et al. 1964; Crick & Grime 1987) and variations in size and placement of resource-acquiring organs such as leaves and roots are of major importance for plant adjustment to resource availability (Bazzaz & Harper 1977; Crick & Grime 1987; Dong et al. 1996). Possibly, the best known functional patterns of morphological plasticity for two essential resources-nutrients and light-involve greater LA relative to plant biomass under low photon flux density (Smith 1982; Reich et al. 1998b; Poorter 2001; Navas & Garnier 2002; King 2003; Steinger et al. 2003). This study also showed greater LA in plants present in moist site where SMC, P and clay are more (as shown by the positive relationships of LA with SMC, P and clay), density of tree species is high and photon flux density is low. Lower LA was observed at dry site where SMC was low and light intensity was high. For example, LA of *A. catechu* was 48 cm^2 at Hathinala and 32 cm^2 at Kotwa, *D. melanoxylon* showed its maximum LA (68 cm^2) at Hathinala and minimum (46 cm^2) at Kotwa, *E. officinalis* showed LA of 238 cm^2 at Hathinala and 185 cm^2 at Kotwa, LA of *L. coromandelica* was 197 cm^2 at Hathinala and 138 cm^2 at Kotwa. These species are the typical species of TDF and have shown high plasticity in LA as well as other FTs.

Table 8. Pearson correlation coefficients between soil moisture content (SMC) and functional traits across species and sites

	RGR	Bio incr	WUEi	A_{max}	Gs_{max}	Chl	LPC	LNC	LCC	SLA	LDMC	RWC	LAI	CD	CC	LA	WSG	BT	HT	
SMC	-0.59***	0.15ns	-0.66***	0.20*	0.68***	0.62***	0.55***	-0.0ns	-0.56***	0.30***	-0.67***	0.52***	0.61***	0.61***	0.59***	0.21*	0.19*	0.23*	0.74***	0.58***
GT	-0.89***	0.54***	-0.32**	0.31**	0.44***	0.31**	0.36***	0.39***	-0.30**	0.08ns	-0.24*	0.15ns	0.84***	0.86***	0.91***	0.44***	0.23*	0.76***	0.93***	
HT	-0.83***	0.44***	-0.36***	0.39***	0.53***	0.46***	0.43***	0.33**	-0.40***	0.20*	-0.38***	0.23*	0.85***	0.89***	0.88***	0.39***	0.29**	0.65***		
BT	-0.61***	0.53***	0.01ns	0.29**	0.14ns	0.09ns	0.11ns	0.51***	0.01ns	-0.09ns	0.01ns	-0.16ns	0.56***	0.56***	0.67***	0.45***	0.16ns			
WSG	-0.17ns	0.00ns	0.22*	0.38***	0.08ns	0.06ns	0.08ns	0.36***	-0.04ns	0.31**	0.13ns	0.22*	0.35***	0.41***	0.29**	-0.43***				
LA	-0.37***	0.44***	-0.28**	-0.06ns	0.21*	0.06ns	0.12ns	0.05ns	-0.04ns	-0.17ns	-0.26**	-0.11ns	0.36***	0.19*	0.25**					
CC	-0.84***	0.39***	-0.35***	0.38***	0.51***	0.44***	0.47***	0.43***	-0.43***	0.26**	-0.18ns	0.24*	0.70***	0.87***						
CD	-0.76***	0.46***	-0.27**	0.41***	0.48***	0.36***	0.39***	0.37***	-0.35***	0.22*	-0.19*	0.22*	0.74***							
LAI	-0.73***	0.41***	-0.26**	0.44***	0.53***	0.41***	0.42***	0.36***	-0.37***	0.41***	-0.28**	0.22*								
RWC	-0.23*	-0.02ns	-0.35***	0.17ns	0.40***	0.43***	0.53***	0.03ns	-0.35***	0.37***	-0.40***									
LDMC	0.28**	-0.19*	0.52***	-0.06ns	-0.45***	-0.49***	-0.37***	0.15ns	0.44***	-0.12ns										
SLA	-0.11ns	-0.08ns	-0.18ns	0.59***	0.57***	0.61***	0.53***	0.23*	-0.59***											
LCC	0.34***	-0.05ns	0.45***	-0.37***	-0.59***	-0.65***	-0.55***	-0.07ns												
LNC	-0.30**	0.42***	0.25**	0.54***	0.15ns	0.13ns	0.19*													
LPC	-0.43***	0.09ns	-0.49***	0.35***	0.63***	0.73***														
Chl	-0.36***	0.07ns	-0.55***	0.40***	0.73***															
Gs_{max}	-0.44***	0.28**	-0.71***	0.52***																
A_{max}	0.21*	0.24**	0.20*																	
WUEi	0.42***	-0.14ns																		
Bio incr	-0.34***																			

Table 8. (Continued)

[ns]$P > 0.05$, [*]$P < 0.05$, [**]$P < 0.01$, [***]$P < 0.001$

HT, height; BT, bark thickness; WSG, wood specific gravity; LA, leaf area; CC, canopy cover; CD, canopy depth; LAI, leaf area index; RWC, leaf relative water content; LDMC, leaf dry matter content; SLA, specific leaf area; LCC, leaf carbon concentration; LNC, leaf nitrogen concentration; LPC, leaf phosphorus concentration; Chl, chlorophyll concentration; Gs_{max}, stomatal conductance; A_{max}, photosynthetic rate; WUEi, leaf water use efficiency; Bio Incr, biomass increment; RGR, relative growth rate. n = 112, [ns]$P > 0.05$, [*]$P < 0.05$, [**]$P < 0.01$, [***]$P < 0.001$.

HT and LAI are important morphological traits which help the plant species to capture light in dense habitats and are a useful measure of competitive advantage, therefore, their values are recorded high in the plant species present in the moist site, Hathinala. When observed in *A. catechu*, its HT was 7 m at Hathinala and 4 m at Kotwa. Similarly its LAI was 11 at Hathinala and 7 at Kotwa. Higher LAI have been attributed to shifts in the allocation of growth from belowground to aboveground in response to increased soil nutrients (Gower et al. 1992) (as observed at Hathinala where organic C, total N and total P in soil are more than other sites and showed significant positive relationships with LAI), or to superior leaf area efficiency (above ground net primary production/LAI) with differences in stand structure (Long & Smith 1990), canopy position (Gilmore & Seymour 1997; Maguire et al. 1998) or site quality (Waring et al. 1980). In the present study this pattern was observed in most of the plant species of the study site. For example in *A. catechu* along with its greater LAI, it's CC (16 m^2) and CD (4 m) were also greater at Hathinala as compared to its value at Kotwa (CC, 12 m^2; CD, 3 m).

It has been reported that greater BT protects the plant from extreme temperature, desiccating winds, herbivory, physical abrasion and fire (Romberger et al. 1992). This reporting is supported by the present study which showed maximum BT in plant species of dry sites as compared to wet ones. These are the sites where disturbance including fire incidences are more.

WSG is positively related to drought resistance in tropical trees (Hacke et al. 2001; Meinzer 2003; Slik 2004; Van Nieuwstadt 2002). This resistance is linked to the fact that high WSG is positively associated with xylem wall enforcement, which reduces cavitation risk due to strong tensions during periods of drought (Hacke et al. 2001). In this study also, WSG was high in the species growing in dry site. For example WSG of *A. catechu* was 0.66 g cm^{-3} at Hathinala and 0.68 g cm^{-3} at Kotwa, *A. latifolia* showed WSG of 0.61 g cm^{-3} at Hathinala and 0.63 g cm^{-3} at Kotwa.

Morphological plasticity, appears to represent a high-cost solution to a change in environment (Bradshaw 1965), whereas, physiological plasticity is usually associated with a change in properties brought about by reversible sub-cellular rearrangements, and represents lower costs and a more rapid response to environmental stress. Nevertheless, physiological adjustments may act as a primary signal that could lead to longer-term responses in morphology. In this study also, physiological plasticity is more compared to morphological plasticity in the related traits. For example LA is related to canopy light interception and photosynthetic efficiency and contributes to carbohydrate metabolism, dry matter accumulation, yield and RGR (Chaturvedi &

Raghubanshi 2011). This study showed significant positive correlation of LA with LAI ($R = 0.36$, $P < 0.001$). Further, LAI was positively correlated with SLA ($R = 0.41$, $P < 0.001$). SLA shows negative correlation with LCC (Ryser & Eek 2000), and positive with RWC (Lichtenthaler et al. 2007), LNC (Niinemets 1999), RGR, Bio Incr, Gs_{max}, biochemical parameters related to photosynthesis, leaf longevity and its palatability (Meziane & Shipley 1999; Weiher et al. 1999). These patterns are also observed in this study (*viz.*, LCC, $R = -0.59$, $P < 0.001$; RWC, $R = 0.37$, $P < 0.001$; LNC, $R = 0.23$, $P < 0.05$; Chl, $R = 0.61$, $P < 0.001$; Gs_{max}, $R = 0.59$, $P < 0.001$; $R = 0.57$, $P < 0.001$).

These traits work in association for better optimization of photosynthesis. According to Santiago et al. (2004), A_{max} tends to decrease with increasing leaf life-span. It also decreases with increasing tree age and size (Niinemets et al. 2009). A positive correlation with Gs_{max} and SLA has been observed by Niinemets et al. (2009) and is shown to be influenced by structure, LNC, Gs_{max} and carboxylation capacity of leaf (Wright et al. 2004). In the present study, most of the morphological traits showed significant positive correlations with A_{max} (HT, $R = 0.39$, $P < 0.001$; BT, $R = 0.29$, $P < 0.01$; WSG, $R = 0.38$, $P < 0.001$; CC, $R = 0.38$, $P < 0.001$; CD, $R = 0.41$, $P < 0.001$; LAI, $R = 0.44$, $P < 0.001$). Among physiological traits, SLA ($R = 0.59$, $P < 0.001$), LNC ($R = 0.54$, $P < 0.001$), LPC ($R = 0.35$, $P < 0.001$), Chl ($R = 0.40$, $P < 0.001$), Gs_{max} ($R = 0.52$, $P < 0.001$), WUEi ($R = 0.20$, $P < 0.05$) and Bio Incr ($R = 0.24$, $P < 0.01$) showed significant positive association with A_{max}.

This study showed that among morphological traits, HT, LA, CC, CD and LAI were positively related with SMC, C, N, P and clay. These soil properties are prevalent in moist sites and plant species present in these sites better utilize the resources by increasing their HT, LA, CC, CD and LAI. On the other hand, BT and WSG have shown positive relations with BD, pH and sand content. These soil properties are greater at dry sites and the morphological traits to which they are positively related help the plant species to tolerate the harsh conditions of the dry sites.

Among the physiological traits, RWC, SLA, LPC, Chl and Gs_{max} were positively related with SMC, C, N, P and clay, whereas, LDMC, LCC, LNC, A_{max} and WUEi showed positive association with BD, pH and sand. Therefore, we can interpret that RWC, SLA, LPC, Chl and Gs_{max} are the important physiological traits for the plant species present in moist habitat which contains high stem density and LDMC, LCC, LNC, A_{max} and WUEi are important for the plant species present in dry habitat where SMC is low. It has also been observed that the two growth traits, i.e., Bio Incr and RGR were positively correlated with BD, pH and sand. Thus, a particular plant species present in dry sites may have greater Bio Incr and RGR than the the same present at moist sites.

Most of the plant species in the study sites are light demanding and highly deciduous. According to Keeling et al. (2008) Bio Incr of tree species is affected by FTs. Physiological and morphological trade-offs imply that shade-tolerant species maximise Bio Incr in the shade due to lower whole-plant light compensation points, whereas, in high light, light-demanding species have faster Bio Incr due to greater A_{max} (Pacala et al. 1994; Walters & Reich 2000). According to Keeling et al. (2008) light demanding species typically have faster foliage turnover, which acts as a significant biomass "drain", and therefore greater biomass production is necessary to maintain sufficient leaf biomass. Fast RGR in these species can be achieved through an efficient uptake and/or use of resources such as water, nutrients and light (Norgren 1996).

The growth rate depends upon the degree of competition, thus its pattern can be controlled to a large degree through spacing (Philip 1994). Increasing stem diameter with increasing the distance between trees is simply a result of exploiting same available below-ground resources (water and nutrient) by less number of trees (Aref et al. 1999). Increasing spacing would also increase biomass of branches, leaves and main roots of trees (Hongtong 1990). Increasing growth rate of diameter with increasing the spacing was also reported by Saatawut & Tularak (1986), Orlic (1987), Vacharangkura (1988) and Effendi & Bachtiar (1991). In the present study, the five study sites have similar soil nutrient content, most of the species except *S. robusta* are light demanding and deciduous species, therefore, their biomass increment and RGR were maximum at the dry sites where tree density was less.

Present study shows that all FTs under study affect RGR directly or indirectly. However, the strength of effect is determined by environmental parameters and in case of TDF soil water availability is the important parameter. Step-wise multiple regression indicates that more than 80% variability in RGR can be explained by CC, LAI, SLA and WUEi alone (Table 9). First three variables represent quantity of photosynthetic surface and last represent water use economy of a species. All these are also significantly modulated by soil moisture availability. Important point to note here is that A_{max} is not an important parameter to determine RGR in TDF where water economy and extended period of leaflessness are critical.

Table 9. Result of step-wise multiple regression to predict RGR on the basis of selected morphological and physiological traits. CC, canopy cover; LAI, leaf area index; SLA, specific leaf area; WUEi, intrinsic water use efficiency

Coefficients[a]						
Model	Unstandardized Coefficients		Standardized Coefficients	T	Sig.	
	B	Std. Error	Beta			
(Constant)	0.057	0.012		4.724	0.000	
CC	-0.003	0.000	-0.586	-9.538	0.000	
LAI	-0.004	0.001	-0.371	-5.880	0.000	
SLA	0.000	0.000	0.224	4.742	0.000	
WUEi	0.001	0.000	0.154	3.360	0.001	

a. Dependent Variable: RGR.

ACKNOWLEDGMENTS

The authors thank Divisional Forest Officer, Renukoot, Sonebhadra, Uttar Pradesh India for granting permission to work in the forest. R.K. Chaturvedi thanks Council of Scientific and Industrial Research for funding support in the form of Senior Research Fellow (Extended).

REFERENCES

Aerts R & Chapin FS (2000) The mineral nutrition of wild plants revisited: a re-evaluation of processes and patterns; *Advances in Ecological Research 30:* 1–67.

Aref IM, El-Juhany LI & Nasroon TH (1999) Pattern of above-ground biomass production and allocation in *Leucaena leucocephala* trees when planted at different spacing; *Saudi Journal of Biological Sciences 6:* 27-34.

Bazzaz FA & Harper JL (1977) Demographic analysis of the growth of *Linum usitatissimum*; *New Phytologist 78:* 193–208.

Bradshaw AD (1965) Evolutionary significance of phenotypic plasticity in plants; *Advances in Genetics 13:* 115–155.

Bradshaw AD, Chadwick MJ, Jowett D & Snaydon RW (1964) Experimental investigations into the mineral nutrition of several grass species; *Journal of Ecology 13:* 665–676.

Bremner JM & Mulvaney CS (1982) Nitrogen-total. In: Page AL, Miller RH & Keeney DR (Eds), Methods of Soil Analysis: Part 2. Chemical and Microbiological Properties. Agronomy Monograph No. 9, second ed. American Society of Agronomy and Soil Science Society of America, Madison, pp. 595–624.

Callahan HS (2005) Using artificial selection to understand plastic plant phenotypes; *Integrative and Comparative Biology 45:* 475–485.

Chaturvedi RK & Raghubanshi AS (2011) *Plant Functional Traits in a Tropical Deciduous Forest: An analysis*; (Lambert Academic Publishing GmbH & Co. KG, Berlin, Germany).

Chaturvedi RK, Raghubanshi AS & Singh JS (2011) Carbon density and accumulation in woody species of tropical dry forest in India; *Forest Ecology and Management 262:* 1576–1588.

Clausen J, Keck DD & Hiesey WM (1940) *Experimental Studies on the Nature of Species I. Effect of Varied Environments on Western North American Plants;* (Carnegie Institute of Washington, Washington (DC). Publication No. 520).

Crick JC & Grime JP (1987) Morphological plasticity and mineral nutrient capture in two herbaceous species of contrasted ecology; *New Phytologist 107:* 403–414.

Curtis JT & Mcintosh RP (1951) An upland forest continuum in the prairie-forest border region of Wisconsin; *Ecology 32:* 476–496.

Dong M, During HJ & Werger MJA (1996) Morphological responses to nutrient availability in four clonal herbs; *Vegetatio 123:* 183–192.

Effendi R & Bachtiar A (1991) The influence of spacing against the diameter and tree height growth of bakau (*Rhizophora apiculata*) in Tembilahan, Riau. Buletin Penelitian Hutan (Indonesia), *Forest Research Bulletin 540:* 35-44.

Gilmore DW & Seymour RS (1997) Crown architecture of *Abies balsamea* from four canopy positions; *Tree Physiology 17:* 71-80.

Givnish TJ (1987) Comparative studies of leaf form: assessing the relative roles of selective pressures and phylogenetic constraints; *New Phytologist 106:*131–160.

Gower ST, Vogt KA & Grier CC (1992) Carbon dynamics of rocky mountain Douglas-fir: influence of water and nutrient availability; *Ecological Monographs 62:* 43–65.

Hacke UG, Sperry JS, Pockman WT, Davis SD & Mcculloh KA (2001) Trends in wood density and structure are linked to prevention of xylem implosion by negative pressure; *Oecologia 126:* 457–461.

Hongtong B (1990) Spacing trials of *Rhizophora apiculata* Sl and *Rhizophora mucronata* Poir. at Amphoe Kantang, Changwat Trang, Bangkok (*M. Sc. Thesis,* Kasetsart University, Bangkok, Thailand).

Huber H (1996) Plasticity of internodes and petioles in prostrate and erect *Potentilla* species; *Functional Ecology 10:* 401–409.

Keeling HC, Baker TR, Vasquez Martinez R, Monteagudo A & Phillips OL (2008) Contrasting patterns of diameter and biomass increment across tree functional groups in Amazonian forests; *Oecologia 158:* 521–534.

King DA (2003) Allocation of above-ground growth is related to light in temperate deciduous saplings; *Functional Ecology 17:* 482–488.

Lewis MC (1972) The physiological significance of variation in leaf structure; *Science Progress (London)* **60:** 25–51.

Lichtenthaler HK, Ač A, Marek M, Kalina J & Urban O (2007) Differences in pigment composition, photosynthetic rates and chlorophyll fluorescence images of sun and shade leaves of four tree species; *Plant Physiology and Biochemistry 45:* 577-588.

Long JN & Smith FW (1990) Determinant of stem wood production in *Pinus contorta* var. *latifolia* forest: the influence of site quality and stand structure; *Journal of Applied Ecology 27:* 847-856.

Maguire DA, Brissette JC & Gu L (1998) Crown architecture and growth efficiency of red spruce in uneven-aged, mixed species stands in Maine; *Canadian Journal of Forest Research 28:* 1233-1240.

Mathieu A, Cournède PH, Letort V, Barthélémy D & de Reffye P (2009) A dynamic model of plant growth with interactions between development and functional mechanisms to study plant structural plasticity related to trophic competition; *Annals of Botany 103:* 1173 – 1186.

Meinzer FC (2003) Functional convergence in plant responses to the environment; *Oecologia 134:* 1–11.

Meziane D & Shipley B (1999) Interacting determinants of specific leaf area in 22 herbaceous species: effects of irradiance and nutrient availability; *Plant, Cell & Environment 22:* 447–459.

Mony C, Thiébaut G, Muller S (2007) Changes in morphological and physiological traits of the freshwater plant Ranunculus peltatus with the phosphorus bioavailability; *Plant Ecology 191:* 109-118.

Navas M-L & Garnier E (2002) Plasticity of whole plant and leaf traits in Rubia peregrina in response to light, nutrient and water availability; *Acta Oecologica 23:* 375–383.

Niinemets Ü (1999) Research review: components of leaf dry mass per area – thickness and density – alter leaf photosynthetic capacity in reverse directions in woody plants; *New Phytologist 144:* 35–47.

Niinemets Ü, Diaz-Espejo A, Flexas J, Galmes J & Warren CR (2009) Role of mesophyll diffusion conductance in constraining potential photosynthetic productivity in the field; *Journal of Experimental Botany 60:* 2249-2270.

Norgren O (1996) Growth analysis of Scots pine and lodge pole pine seedlings; *Forest Ecology and Management 86:* 15–26.

Olsen SR & Sommers LE (1982) Phosphorus, chemical and microbiological properties. In: Page AL, Miller RH & Keeney DR (Eds), Methods of Soil Analysis: Part 2 Agronomy Monograph No. 9, Second ed. American Society of Agronomy and Soil Science Society of America, Madison, pp. 403–430.

Orlic S (1987) Influence of the spacing in planting of Norway spruce (*Picea abies* Karst) on its growth in the Plesivicko Prigorje area (Piadmont Region, Yugoslavia); *Sumarski-list 61:* 221-247.

Pacala SW, Canham CD, Silander JA & Kobe RK (1994) Sapling growth as a function of resources in a north temperate forest; *Canadian Journal of Forest Research 24:* 2172–2183.

Philip MS (1994) *Measuring Trees and Forests* (Cab International, Wallingford, Oxon, United Kingdom).

Poorter L (2001) Light-dependent changes in biomass allocation and their importance for growth of rainforest tree species; *Functional Ecology 15:* 113–123.

Poorter L (1999) Growth response of 15 rain-forest tree species to a light gradient: the relative importance of morphological and physiological traits; *Functional Ecology 13:* 396–410.

Poorter L, Wright SJ, Paz H, Ackerly DD, Condit R, Ibarra-Manríquez G, Harms KE, Licona JC, Martínez-Ramos M, Mazer SJ, Muller-Landau HC, Peña-Claros M, Webb CO & Wright IJ (2008) Are functional traits good predictors of demographic rates? Evidence from five neotropical forests; *Ecology 89:*1908-1920.

Reich PB & Oleksyn J (2004) Global patterns of plant leaf N and P in relation to temperature and latitude; *Proceedings of the National Academy of Sciences, USA 101:* 11001–11006.

Reich PB, Tjoelker MG, Walters MB, Vanderkelin DW & Buschera S (1998) Close association of RGR, leaf and root morphology, seed mass and shade tolerance in seedlings of nine boreal tree species grown in high and low light; *Functional Ecology 12:* 327–338.

Robinson D & Rorison IH (1988) Plasticity in grass species in relation to nitrogen supply; *Functional Ecology 2:* 249–257.

Romberger JA, Hejnowicz Z & Hill JF (1992) *Plant Structure: Function and Development* (Springer-Verlag, Berlin).

Ryser P & Eek L (2000) Consequences of phenotypic plasticity vs. interespecific differences in leaf and root traits for acquisition of aboveground and belowground resources; *American Journal of Botany 87:* 402-411.

Saatawut P & Tularak A (1986) *Spacing Trail of Eucalyptus camaldulinsis Dehnh,* Research Report of Silvicultural (Royal Forestry Department, Bangkok, Thailand) pp 121-126.

Santiago LS, Goldstein G, Meinzer FC, Fisher JB, Machado K, Woodruff D & Jones T (2004) Leaf photosynthetic traits scale with hydraulic conductivity and wood density in Panamanian forest canopy trees; *Oecologia 140:* 543–550.

Schlichting CD (1986) The evolution of phenotypic plasticity in plants; *Annual Review of Ecology and Systematics 17:* 667–693.

Schlichting CD (1989) Phenotypic plasticity in *Phlox*: II Plasticity of character correlations; *Oecologia 78:* 496–501.

Sheldrick BH & Wang C (1993) Particle-size distribution. In: Carter, M.R. (Eds.), Soil Sampling and Methods of Analysis. Canadian Society of Soil Science. Lewis Publishers, Ann Arbor, Michigan, USA, pp. 499–511.

Singh JS, Singh KD (2011) Silviculture of Dry Deciduous Forests, India. In: Günter S, Weber M, Stimm B, Mosandl R, ed. Silviculture in the Tropics. Springer-Verlag Berlin, Heidelberg. pp 273-283.

Slik JWF (2004) El Nino droughts and their effects on tree species composition and diversity in tropical rain forests; *Oecologia 141:* 114–120.

Smith H (1982) Light quality, photoreception and plant strategy; *Annual Review of Plant Physiology 33:* 481–518.

Steinger T, Roy BA, Stanton ML (2003) Evolution in stressful environments. II: adaptive value and costs of plasticity in response to low light in Sinapis arvensis; *Journal of Evolutionary Biology 16:* 313–323.

Sultan SE (1995) Phenotypic plasticity and plant adaptation; *Acta Botanica Neerlandica 44:* 363–383.

Sultan SE & Bazzaz FA (1993) Phenotypic plasticity in *Polygonum persicaria*. III. The evolution of ecological breadth for nutrient environment; *Evolution 47:* 1050–1071.

Sultan SE (2000) Phenotypic plasticity for plant development, function and life history; *Trends in Plant Science 5:* 537–542.

Vacharangkura T (1988) Production of 4-year-old and eight planting spaces of *Eucalyptus camaldulensis* Dehnh. Plantations at Changwat SiSaket, Thailand (*M. Sc. Thesis,* Kasetsart University, Bangkok, Thailand).

Valladares F, Wright SJ, Lasso E, Kitajima K & Pearcy RW (2000) Plastic phenotypic response to light of 16 congeneric shrubs from a Panamanian rainforest; *Ecology 81:* 1925–1936.

Van Nieuwstadt MGL (2002) *Trial by fire: postfire development of a tropical dipterocarp forest* (Ph.D. thesis, Utrecht University).

Walkley A & Black IA (1934) An examination of the Degtjareff method for determining soil organic matter, and a proposed modification of the chromic acid titration method; *Soil Science 37:* 29–38.

Walters MB & Reich PB (2000) Trade-offs in low-light CO_2 exchange: a component of variation in shade tolerance among cold temperate tree seedlings; *Functional Ecology 14:*155–165.

Waring RH, Thies WG & Muscato D (1980) Stem growth per unit leaf area: a measure of tree vigor; *Forest Science 26:* 112-117.

Weiher F, Van der Werf A, Thompson K, Roderick M, Garnier E & Eriksson O (1999) Challenging Theophrastus: a common core list of plant traits for functional ecology; *Journal of Vegetation Science 10*: 609–620.

Westoby M, Falster DS, Moles AT, Vesk PA & Wright IJ (2002) Plant ecological strategies: some leading dimensions of variation between species; *Annual Review of Ecology and Systematics 33:* 125–159.

Winn AA & Evans AS (1991) Variation among populations of *Prunella vulgaris* L. in plastic responses to light; *Functional Ecology 5:* 562–571.

Winn AA (1996a) Adaptation to fine-grained environmental variation: an analysis of within-individual leaf variation in an annual plant; *Evolution 50:* 1111–1118.

Winn AA (1996b) The contributions of programmed developmental change and phenotypic plasticity to within-individual variation in leaf traits in *Dicerandra linearifolia*; *Journal of Evolutionary Biology 9:* 737–752.

Winn AA (1999) The functional significance and fitness consequences of heterophylli; *International Journal of Plant Sciences 160:* S113–S121.

Wright IJ, Ackerly DD, Bongers F, Harms KE, Ibarra-Manriquez G, Martinez-Ramos M, Mazer SJ, Muller-Landau HC, Paz H, Pitman NCA, Poorter L,

Silman MR, Vriesendorp CF, Webb CO, Westoby M & Wright SJ (2007) Relationships Among Ecologically Important Dimensions of Plant Trait Variation in Seven Neotropical Forests; *Annals of Botany 99:* 1003–1015.

Wright IJ, Reich PB, Cornelissen JHC, Falster DS, Garnier E, Hikosaka K, Lamont BB, Lee W, Oleksyn J, Osada N, Poorter H, Villar R, Warton DI & Westoby M (2005) Assessing the generality of global leaf trait relationships; *New Phytologist 166:* 485–496.

Wright IJ, Reich PB, Westoby M, Ackerly DD, Baruch Z, Bongers F, Cavender-Bares J, Chapin T, Cornelissen JHC, Diemer M, Flexas J, Garnier E, Groom PK, Gulias J, Hikosaka K, Lamont BB, Lee T, Lee, W, Lusk C, Midgley JJ, Navas M-L, Niinemets Ü, Oleksyn J, Osada N, Poorter H, Poot P, Prior L, Pyankov V I, Roumet C, Thomas SC, Tjoelker MG, Veneklaas EJ & Villar R (2004) The worldwide leaf economics spectrum; *Nature 428:* 821-827.

In: Phenotypic Plasticity

ISBN: 978-1-62618-404-6

Editors: J. Valentino and P. Harrelson

© 2013 Nova Science Publishers, Inc.

Chapter 3

UNDERSTANDING THE MECHANISMS OF CHEMICAL COMMUNICATION AND ITS CONSEQUENCES ON AQUATIC COMMUNITIES USING THE MODEL FRESHWATER CRUSTACEAN DAPHNIA

Linda C. Weiss, Florian Leese and Ralph Tollrian*
Department of Animal Ecology Evolution and Biodiversity,
Ruhr-University Bochum, Bochum, Germany

ABSTRACT

Phenotypic plasticity is defined as the ability of an organism with a given genotype to respond to changing environmental conditions by undergoing a phenotypic adaptation. Predator-induced defences are one form of phenotypic plasticity. Only when needed i.e. in the presence of a predator, organisms build defensive structures (e.g. crests, thorns or spines) and/or behaviours in a cost-benefit efficient manner. In aquatic ecosystems these phenotypic changes are frequently triggered by predator-specific chemical cues (kairomones). This kind of chemical information transfer has a major impact on food-web interactions as inducible defences dampen predator-prey oscillations. This stabilizes populations and increases their persistence.

---

* Corresponding author: Linda Weiss; linda.weiss@rub.de.

The ecology and evolution of these chemically mediated anti-predator defences have been intensely studied especially in the freshwater crustacean *Daphnia*. Together with the sequencing of the genome the freshwater crustacean *Daphnia* has been established as a new model organism.

The *Daphnia* genome holds remarkable aspects that are hypothesized to allow for the observed flexibility towards environmental changes. An enormous number of at least 30,000 genes, several of which are members of recently diverged gene families, characterize the *Daphnia* genome. Such exceptional and almost unique features call for more detailed studies. Besides the study of the genome evolution in *Daphnia*, the determination of the functional genomic and physiological mechanisms that enable *Daphnia* to adapt to changing environmental conditions are of particular interest. These mechanisms include processes suitable for the detection of environmental conditions, and features that allow an effective phenotypic alteration.

This chapter will review information on the cellular and neuronal mechanisms involved in the physiology underlying the formation of inducible defences. This will be discussed together with information available from genome data.

We will consider state of the art research and suggest future directions leading to a more detailed understanding of how phenotypes adapt to changing environments.

THE ECOLOGY OF PHENOTYPIC PLASTICITY

Phenotypic plasticity, describes the ability of an organism with a given genotype to develop an adaptive phenotype meeting current environmental conditions (West-Eberhard 2003). Thereby an organism's fitness is adjusted to the present environmental state, which allows population persistence and stability (Mougi & Kishida 2009; Verschoor et al. 2004). Phenotypic plasticity is common among living things. Essentially, phenotypic plasticity derives from the instability of natural environments. Organisms counter these different environmental changes by either between- or within-generation variation (Langerhans & DeWitt 2002). Between-generation variation has been a primary focus of evolutionary biologists and is based on natural selection acting on heritable variation caused by mutation, recombination and/ or gene flow. Also within-generation variation is on a genetic background, but it rather depends on mechanisms including differential gene expression, and potentially alternate splicing, transcription and/ or translation (Langerhans & DeWitt

2002). This allows individuals to adjust to environmental variation in real time.

Environmental adaptations are classified in different categories including any type of physiological and behavioural adaptation. E.g. seasonal polyphenism of wing coloration in butterflies is an example for developmental plasticity (Prudic et al. 2011). But also shifts from clonal to sexual reproduction triggered by photoperiodic cues can be observed in aphids (Le Trionnaire et al 2008). By using the same set of genes aphids are able to make use of these two extremely different reproductive modes. The photoperiodic length is measured in the cephalic region and the signal is transduced to the target cells located in the ovaries (Le Trionnaire et al 2008). All environmental information is generally received by sensory neurons. The number of chemosensilla on the antenna can be altered in grasshoppers in response to the number of plant chemicals that are encountered (Chapman & Lee 1991). Frequent stimulation of such neurons may render the connectivity of neuronal circuits within the nervous system, which can result in alternate behaviours (Montarolo et al. 1986). Therefore, neuronal plasticity and behavioural plasticity are tightly connected and represent further examples of phenotypic plasticity. An astonishing example of phenotypic plasticity can be seen in the defensive trait formation in a prey organism as response to the presence of predator species.

INDUCIBLE DEFENCES

Inducible defence strategies are developed by prey species upon an increased risk of predation. These defences are frequently predator specific and serve as a protection against an increased predation risk. Besides having the ability to determine the presence of a predator most prey species can determine predator densities evoking concentration-dependent responses. Meaning that the predator density is positively correlated with the risks of being attacked and the response is generally dependent on the predator's population size. This increases prey survival chances and thus the fitness of organisms under this specific environmental condition. At the same time inducible defences incur costs (Tollrian & Harvell 1999, Riessen 2012) and therefore traits are only developed when they are needed and costs are saved when they are not essential. Determination of predator density, impact and adequate predator identification is crucial for the survival of prey organisms. Only accurate assessment of the predation risk allows a forceful and cost-benefit optimised development of defences. Whereas it is beneficial to remain small and transparent in the presence of visually hunting predators e.g. fish, it is of advantage to grow large or even bulky when encountering threats from gape-limited predators e.g. insect larvae (Tollrian 1993), fish (Tollrian 1994; Figure 1) or the ancient predator *Triops* spec. (Rabus & Laforsch 2011).

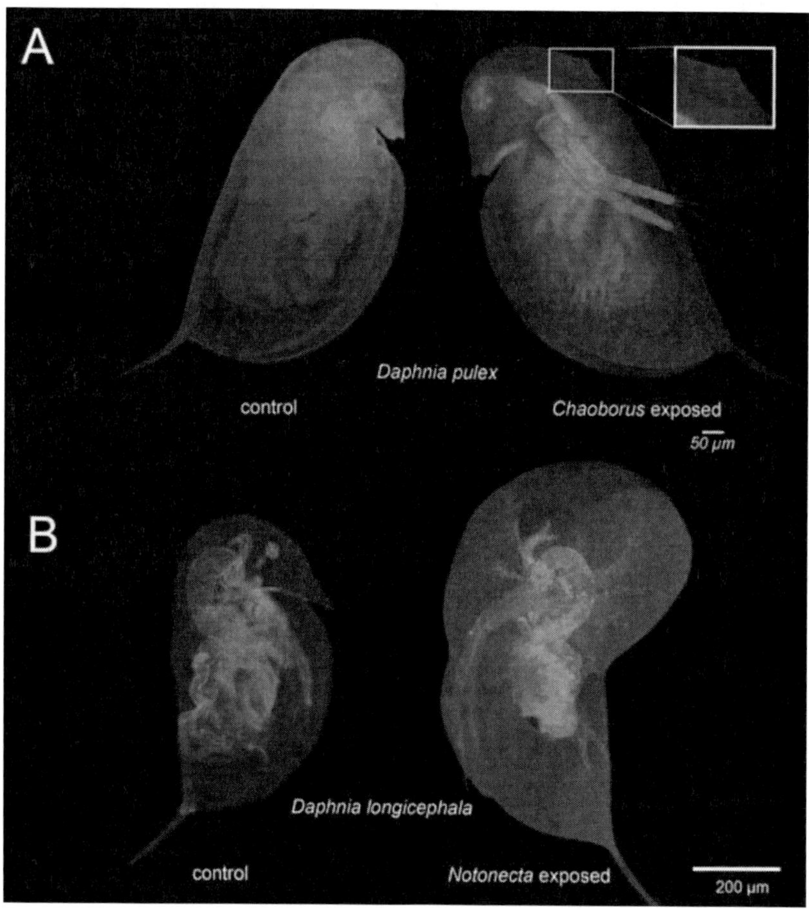

Figure 1. Inducible morphological defences in *Daphnia*. A: Unexposed (left) induced morphotype (right) with developed neckteeth in *D. pulex* as a defence against phantom midge larvae (*Chaoborus flavicans*, Diptera).B: Unexposed (left) and induced morphotype (right) with a developed crest in *D. longicephala* as a defence against the backswimmer (*Notonecta glauca*, Heteroptera).

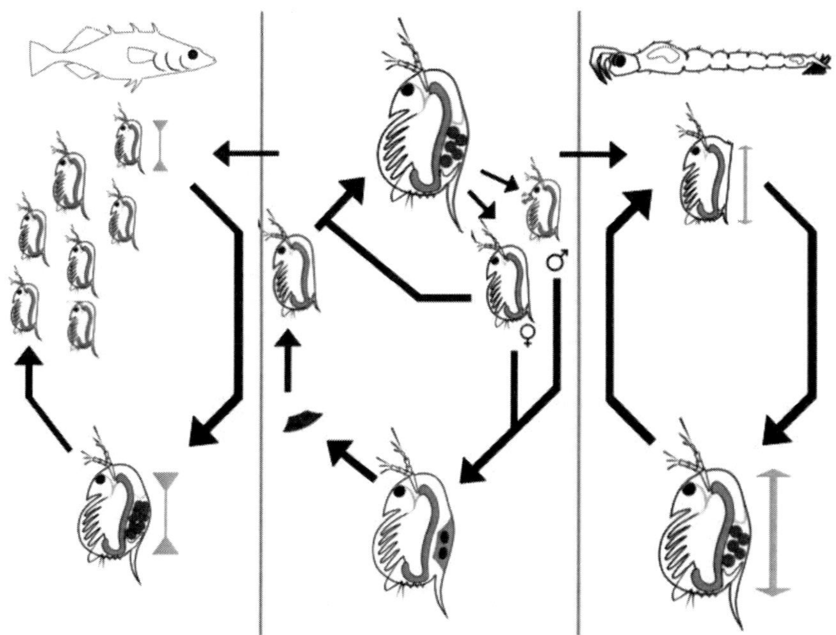

Figure 2. Reproduction cycle and inducible defence strategies in *Daphnia pulex*. In the absence of a predator stress *Daphnia* reproduce through parthenogenesis resulting in the production of female clones. During changing environmental conditions such as change in temperature, *Daphnia* are switch to sexual reproduction and produces male offspring. Fertilization leads to resting egg production. From resting eggs, new female offspring hatch. In the presence of the predatory *Chaoborus* larvae (right panel) *Daphnia pulex* forms specialized morphological spines in the neck region, so called neckteeth. In addition reproduction is traded with somatic growth resulting in an increased body size but a decreased number of offspring. Under fish predation (left panel) resources are allocated from somatic growth to reproduction, resulting in earlier sexual maturity at a smaller size and an increased number of smaller offspring.

Anti-predatory traits can be observed in forms of 1.) behavioural adaptations (Pijanowska & Kowalczewski 1997; Pijanowska et al. 2006; Mathis et al. 2008), 2.) life-history shifts (Kiesecker et al. 2002) and/ or 3.) morphological defences (Tollrian & Harvell 1999) (Kiesecker et al. 2002). 1.) Behavioural responses such as reduced foraging can be observed in specimen exposed to an increased predation risk in predator presence. Thereby prey species might reduce chances of predator encounter at the cost of nutrition uptake (Lampert 1989; Dodson et al. 1997). 2.) In life-history shifts somatic growth is traded with reproduction and vice versa (Ball & Baker 1996; Boersma et al. 1998; Crowl, & Covich 1990; Black 1993). Against visual predators prey species that remain small and thus possibly unnoticed, produce an increased number of offspring. Gape-limited predation can be countered by growing large at the cost of reproduction. 3.) Some species can even develop specialized defensive morphological traits. If captured these morphological changes hinder the predator's grasp and thereby facilitate prey escape (Beckerman et al. 2010). Such defensive morphological peculiarities are widespread in plants and the animal kingdom. They can show e.g. in forms of spines (Tollrian 1994), thorns (Petrusek et al. 2009), neckteeth (Tollrian 1993, Figure 2), crests (Grant & Bayly 1981; Barry 1994; Figure 2), thickened shells (Bourdeau 2010) but even altered colorations ((Price et al. 2003) reviewed in Weiss et al. 2012).

Especially in laboratory experiments, measuring morphological defences is advantageous as their numerical quantification is facilitated in comparison to e.g. behavioural defences.

The Importance of Inducible Defences and their Ecological Implications

Although it is obvious that inducible defences can affect predator-prey relationships, competitive interactions and potentially ecological processes and ecosystem functions, real studies about population effects are rare. Theoretical models describe that populations are stabilised based on decreased predator-prey oscillations (Verschoor et al. 2004; Mougi, & Kishida 2009). With an increasing predator population, kairomones accumulate and in a concentration-dependent manner defences are developed in prey species increasing survival chances. These complicate predator population growth due to higher foraging costs. As chances of outgrowing predator populations become more unlikely, oscillation extremes are reduced and populations are stabilised. However, in order to be advantageous any inducible defence strategy needs to be active in due time (reviewed in Laforsch & Tollrian 2009). The lengths of such lag times vary between strategies. Whereas behavioural responses have short lag times as they can be quickly turned on and off, morphological defences or life-history shifts require longer lag times compared. It appears that morphological defences, being more costive than behavioural defences, must be more effective than behavioural defences. The higher efficiency thus balances the greater costs.

Presumably any kind of abiotic or biotic factor can evolve to serve as a stimulating cue for induced plasticity. Inducible defences, just as plasticity, are manifested on a physiological background that is dependent on a given genome. All phenotypic changes require an adaptive biochemistry, morphology, development and/ or behaviour. Subsequently, the resulting changes may be precisely integrated upon a predator threat in order to form a highly adaptive phenotype.

Numerous mechanisms including alternate genome transcription, translation, and possibly alternate protein folding, must consequently result in enzyme and hormonal regulation, producing local and systemic responses that accomplish plasticity. Moreover, the timing, specificity and speed of plastic responses are critical to their adaptive value.

In the following, we review the first insights that have been obtained in the understanding of inducible defence development in *Daphnia*.

Inducible Defences in *Daphnia*

Predator-specific Cues

The waterflea *Daphnia* spp. has a particularly long history of ecological, ecotoxicological and evolutionary research. *Daphnia's* high P content makes it a particularly nutritious food resource (Elser et al. 2000). In this regard, it is of high ecological relevance as it forms the trophic link between primary producers and higher trophic levels in limnetic ecosystems. In this freshwater crustacean we find the ability to specifically adapt to environmental changes including various predation risks. *Daphnia* forms behavioural (Dodson et al. 1997), morphological (Tollrian 1993) and life history defences (Stibor 1992) in response to chemical cues from vertebrate and invertebrate predators. These so-called kairomones are defined as chemical cues in the interspecific information transfer, that are beneficial to the receiver but not to the sender. Kairomones released by the phantom midge larvae *Chaoborus* (Diptera) e.g. induce neckteeth (Dodson et al. 1997; Tollrian 1993) while kairomones released from fish e.g. *Gasterosteus aculeatus* induce changes in life-history parameters (Pijanowska 1992). Furthermore, behavioural responses such as diel vertical migration (DVM) are induced by fish specific cues (Lampert 1989). As fishes are visually hunting predators *Daphnia* avoid fish encounters by remaining in cooler and darker water strata during the day and migrate to warm and nutritional water strata during the night.

Unfortunately, the chemical nature of many kairomones has not been identified and this is not just limited to *Daphnia* and its predators. Knowledge of the chemical identity of kairomones is essential in order to determine why these substances have evolved to become induction factors in prey species. Moreover, knowing the chemical nature of the kairomone helps to precisely evaluate the underlying signal transduction cascade in prey species, providing knowledge of how predator-information is perceived, decoded and transformed into a physiological response. In the following, we summarize the present knowledge of predator detection in *Daphnia*.

Predator Detection

Animals evaluate and interact with their environment using ancient chemo-sensory modalities including taste and smell (Vosshall 2000; Galizia & Szyszka 2008). The high sensitivity of *Daphnia* towards a diversity of chemical stimuli renders it an ideal species for ecotoxicological research. The presence of a repertoire of 58 chemoreceptors sequences presumably mediates the many chemoreceptive abilities in waterfleas (Peñalva-Arana et al. 2009). These 58 chemoreceptors that cluster in 3 distinctive superfamilies share sequence homology with insect gustatory receptors (Grs). Astonishingly, no genes encoding proteins similar to the insect odorant receptors (Ors) were found, which might indicate a quite recent expansion of this gene lineage concomitant with the evolution of hexapods (Peñalva-Arana et al. 2009). Yet, it is still questionable whether *Daphnia* only relies on these 58 Grs. Chemo-receptor genes seem to have evolved very independently between taxa, i.e. there are no sequence homologies between vertebrates and arthropods. Therefore, it is likely that *Daphnia* also possess their own set of genes coding for yet undescribed chemoreceptors sharing no homologues with hexapods or vertebrates.

The location of these chemoreceptors yet remains to be determined. However, as chemical stimuli in *Daphnia* are thought to be foremost perceived by the sensory cells of the first antenna (antennule), receptors are most likely to be located here. The antennule itself is short in female cladocerans and longer in males (Hallberg et al. 1992). This remarkable characteristic of males is probably responsible for the detection of female mating partners (personal observation). Furthermore, a higher sensitivity towards predator cues can be observed in males. Most scientific research however was performed using females, as their natural occurrence is predominant. In females, nine aesthetascs are found on the antennule. The aesthetascs are thought to represent the specific olfactory receptors in crustaceans (Gnatzy et al. 1984). Hallberg et al. (1992) investigated the aesthetascs of *D. magna* and *D. longispina*, whereas Weiss et al. (2012b) investigated the aestethascs of *D. pulex, D. longicephala* and *D. lumholtzi*. Both described cylindrical cells and referred to them as sensory cells. These bipolar cells innervate the lumen of the aesthetascs and send projections into the deutocerebrum (Hallberg et al. 1992; Weiss et al. 2012b). It is speculated that these cells serve as a first signal integration of olfactory sensation (Weiss et al. 2012b).

The so-called olfactory centres in the central nervous system provide a basis for an integration of chemical stimuli (Atema & Stenzler 1977). The deutocerebrum in branchiopods receives a mixture of chemosensory and mechanosensory input from the antennulae (Harzsch 2006; Fritsch & Richter 2010) and is therefore suggested to be an olfactory centre of the central nervous system in *Daphnia*. In malacostracans (Schmidt & Ache 1992) as well as hexapods and vertebrates (Buck 1996), olfactory glomeruli allow for discriminative odour coding (Eisthen 2002). However, up to date the presence of olfactory glomeruli in *Daphnia* has not been proven. A detailed morphological description of the *Daphnia* olfactory system is essential for the understanding of how the olfactory environment is transformed in neuronal stimuli (Weiss et al. 2012b).

Based on the above described observations, *Daphnia* must have a nervous system that is not only capable to detect and determine the strength of the present predatory risk, but also has the potential to identify the predator allowing the activation of adequate defences. Recent findings describe the neurophysiology underlying different predator cues (Weiss et al. 2012a). Inducible morphological defences resulting from *Chaoborus* specific predator cues were shown to be bound to a neuronal pathway involving cholinergic stimulation. Life history shifts resulting from fish specific cues however are not cholinergically transmitted. Cholinergic stimulation or inhibition had no impact on fish kairomone dependent responses. Thus cholinergic pathways are unlikely to be involved in the perception of fish cues followed by life history shifts. However, GABA receptor stimulation inhibits fish kairomone induced life history shifts. First of all, this provided strong evidence that the nervous system is involved in the physiological pathways underlying the development of fish-dependent life-history responses. In addition, life history parameters induced by fish cues are under the neurophysiological control of GABA. This study showed that inducible defences against *Chaoborus* and fish depend on uncoupled signal transmission systems that act independently. This should be advantageous as it allows for mixed responses, which may have the potential to defend the organism from several predation pressures simultaneously. However, future investigations need to address the effect of two predators present simultaneously. The neurotransmitters applied in these experiments were able to modulate the kairomone response. However, without the kairomone stimulus, neurotransmitters were not able to elicit the development of inducible defences. Thus cholinergic and gabaergic stimulations alone do not activate the defence directly.

These results suggest a pathway with multiple steps, which requires some kind of additional neuronal or endocrine transmission (Weiss et al. 2012a). Recent findings support the involvement of developmental hormones (Miyakawa et al. 2010; Oda et al. 2011).

Developmental Mechanisms

As described above, predator detection appears to be a sequence of biological reactions involving neuronal components, which are converted into an endocrine response. Juvenile hormones represent a group of sesquiterpenoids that regulate many physiological aspects including development, reproduction, diapause and polyphenism (Miyakawa et al. 2010). Both hormones are predominantly known from insect development but recent advances have shown that they are also involved in the formation of inducible morphological defences in *Daphnia*. Elongation of helmets in *D. galeata* was obtained by exposure to one of the juvenile hormone JH mimicking pesticides fenoxycarb (Oda et al. 2011). Furthermore, in *D. pulex* JH pathway genes showed higher expression preceding neckteeth expression, i.e. juvenile hormone acid methyltransferase (JHAMT) mediating the final step of JH synthesis and methoprene-tolerant Met encoding a candidate receptor for juvenile hormone (Miyakawa et al. 2010).

Besides juvenile hormones, the insulin pathway also appears to be involved in the development of neckteeth expression in *D. pulex* (Miyakawa et al. 2010). The insulin pathway is important for the regulation of a variety of developmental processes including body-size (Nijhout 2003; Shingleton et al 2005). The InR gene encodes an insulin or insulin-like growth factor receptor, and the IRS-1 gene encodes an element directly interacting with InR. It has been suggested that the crosstalk between the JH and insulin-signalling pathways is responsible for the expression of morphogenetic factors in the development of beetle horns (Nijhout 2003; Emlen et al. 2006; Miyakawa et al. 2010).

At present our understanding of how *Daphnia* is capable to detect and adjust to such a wide variety of environmental threats is at the very beginning. In order to determine how this small crustacean is capable to adjust to so many hazards, the recently sequenced genome opens new research possibilities.

Daphnia Genome Features

Novel Insights into an Old Ecological Scheme: Genome Research Revives Phenotypic Plasticity!

Many organisms have the ability to specifically adapt to changing environmental conditions. As noted above such an adaptation can either occur by means of between-generation variation or within-generation variation (Langerhans & DeWitt 2002). In general, both processes are genetic. However, the level of genetic processing is essentially different. Between-generation variation can be influenced by natural selection, leading to changes in the gene pool of populations. This gradual and comparatively slow process of genetic variation within a population is thus rather dependent on the genome sequence itself as well as the heritability of genes.

In comparison, within-generation variation involves phenotypic plasticity as a direct response. Phenotypic plasticity is a relatively fast process because no change in the gene pool is required. The involved processes hypothetically include differential gene expression (Fitzpatrick et al. 2005; Moczek & Rose 2009), alternative splicing (Marden 2008), gene-gene communication (Colebourne 2011, please see below), transcription (True & Lindquist 2004), translation (True & Lindquist 2004) and/ or differential protein folding (Sangster & Queistch 2005). Consequently, within-generation variation is based on immediate adaptive processes whose complexity yet remains to be fully understood.

In this regard the freshwater crustacean *Daphnia* is of especial interest. In comparison to other species *Daphnia* appears to be extremely plastic as it can adapt its phcnotype to many environmental scenarios including abiotic and biotic challenges. Investigation of its genetic features will allow determination of the mechanisms responsible for this high degree of plasticity. Together with a transparent carapace, clonal reproductive mode (Figure 2), easy culturing, short generation times, and small size which allows to maintain many clonal strains within one laboratory, *Daphnia* has become a popular model organism. In addition, dormant egg banks are used as historical archives revealing information about adaptations to past conditions in freshwater ecosystems.

The Genetic Fundament of Phenotypic Plasticity in *Daphnia*

In comparison to other genomes the *Daphnia* genome is remarkable in many ways.

Table 1. List of annotated and differentially expressed genes, determined using qPCR

Predator	Prey	Developmental stage	Differentially expressed gene	Gene-ID	Expression level	Homolog Functions in other species	Author
Chaoborus	*Daphnia pulex*	2nd juvenile instar	matrix metallopro-teinase (MMP)	Dappu-303491	up-regulated	extracellular proteases that play important roles in cell-cell signaling processes	Spanier et al. 2010
Chaoborus	*Daphnia pulex*	2nd juvenile instar	glyceraldehyde-3-phosphate dehydrogenase (GAPDH)	Dappu-302823	down-regulated		Spanier et al. 2010
Chaoborus	*Daphnia pulex*	2nd juvenile instar	Alpha-tubulin (aTub)	Dappu-301837	down-regulated		Spanier et al. 2010
Chaoborus	*Daphnia pulex*	1st juvenile instar	juvenile hormone acid methyltrans-ferase (JHMT)	Dappu-300180	up-regulated	encodes the methyltransfe-rase mediating the final step of juvenile hormone synthesis	Miyakawa et al. 2010
Chaoborus	*Daphnia pulex*	1st juvenile instar	Extradenticle (exd)	Dappu-219790	up-regulated	homeobox transcription factor	Miyakawa et al. 2010
Chaoborus	*Daphnia pulex*	1st juvenile instar	Escargot (esg)	Dappu-50534	up-regulated	homeobox transcription factor	Miyakawa et al. 2010
Chaoborus	*Daphnia pulex*	1st juvenile instar	Hox Gene (Hox3)	Dappu-9456	up-regulated	member of the Hox cluster	Miyakawa et al. 2010
Chaoborus	*Daphnia pulex*	1st juvenile instar	insulin-like receptor (InR)	Dappu-270048	up-regulated	Insulin-like growth factor receptor	Miyakawa et al. 2010
Chaoborus	*Daphnia pulex*	1st juvenile instar	Methoprene-tolerant (Met)	Dappu-247693	up-regulated	candidate receptor for juvenile hormone	Miyakawa et al. 2010
Chaoborus	*Daphnia pulex*	1st juvenile instar	insulin receptor substrate-1 (IRS-1)	Dappu-52188 Dappu-304473	up-regulated	Element interacting with Insulin receptor	Miyakawa et al. 2010
Onchorynchus, Salmo, Salvelinus	*D. melanica*	matured	Ebony	n.a.	Up-regulated	Melanin gene	Scosville et al. 2012
(Onchory-nchus, Salmo, Salvelinus	*D. melanica*	matured	DDC	n.a.	Up-regulated	Melanin gene	Scosville et al. 2010

From this rather short list (tab. 1) it becomes evident that higher throughput analysing methods such as RNAseq are required.

In addition, currently the function of the differentially expressed genes is only deduced by homology searches and the respective function in other species. It is therefore essential to, not only determine the differentially expressed genes, but also to describe a potentially tissue specific-localization which will advance our understanding of gene/ protein functions that affect the relationship between genotype and phenotype.

Besides changes of the genome and/ or the gene expression level, that have just been described, plasticity changes could also occur at higher levels including, splicing, translation and protein folding. I. e., it has to be considered that the mRNA is only an intermediate product that advances the production of proteins. However, protein abundance is not necessarily correlated to the amount of the corresponding mRNA transcript (Fröhlich et al. 2009).

Information deduced from genomic data therefore not necessarily reflects the protein reality.

Peptidomics

As described above, molecular mechanism involved in the adaptation to an environment cannot be exclusively deduced from genomic or transcriptional data. Comprehensive datasets addressing the protein level, therefore, are indispensable for a functional characterization of a specific trait. A first step in this regard was performed through, *Daphnia* genome mining. Genome mining represents a valuable tool for the identification of peptides. From identified gene regions Gard (Gard et al. 2009) deduced precursor proteins and predicted the mature peptides. In this study twenty-four peptide-encoding genes were identified, including ones predicted to produce members of the A-type allatostatin, B-type allatostatin, C-type allatostatin, allatotropin (ATR), bursicon a, bursicon b, calcitonin-like diuretic hormone, corazonin, crustacean cardioactive peptide, crustacean hyperglycemic hormone, ecdysis-triggering hormone, eclosion hormone (EH), insulin-like peptide (ILP), molt-inhibiting hormone, neuropeptide F, orcokinin (two genes), pigment-dispersing hormone, proctolin, red pigment concentrating hormone/adipokinetic hormone (RPCH/AKH), short neuropeptide F, SIFamide, sulfakinin, and tachykinin-related peptide (TRP) families/ subfamilies. In total, 96 peptides were predicted from these genes. In future studies it is essential to apply such proteome approaches in order to show differences and similarities in protein expression in Daphnia faced to vertebrate and invertebrate predators. This will also help to discover the proteins and pathways controlling different defensive strategies.

Outlook

The freshwater crustacean *Daphnia* has been shown to be an ideal model species for the investigation of ecological questions. However, investigations on the molecular ecology of the keystone species are just beginning. In this regard the sequencing of the genome is one major hurdle just being accomplished. However, it is now essential to decode the genome's information and investigate the genome structure, function and expression in the light of selection. This will further our understanding of genetic regulation and its physiological outcome in organisms that cope with a diversity of environmental stressors. In this regard, research has to be extended by the investigation of epigenetics, i.e. the non-genetic mechanisms that regulate cellular gene expression to provide a complete picture of the genetics of adaptation.

On top of this, the understanding of the adaptive physiology of *Daphnia* needs to be furthered. This also includes the investigation of the cellular and neuronal mechanisms of phenotypic plasticity as the neurophysiology represents the first step, which is transformed in a physiological response by differential gene regulation. Furthermore, in order to obtain a complete understanding of predator-induced defences the chemical nature of the different kairomones are essential.

REFERENCES

Atema, J. & Stenzler, D., 1977, Alarm substance of the marine mud snail, *Nassarius obsoletus*: biological characterization and possible evolution, *Journal of Chemical Ecology*, 3(2), pp. 173-87.

Ball, S.L. & Baker, R.L., 1996, Predator-Induced Life History Changes: Antipredator Behavior Costs or Facultative Life History Shifts? *Ecology*, 77(4), pp. 1116-24.

Barry, M.J., 1994, The Costs of Crest Induction for *Daphnia carinata*, *Oecologia*, 97(2), pp. 278-88.

Beckerman, A.P., Rodgers, G.M. & Dennis, S.R., 2010, The reaction norm of size and age at maturity under multiple predator risk, *The Journal of Animal Ecology*, 79(5), pp. 1069-76.

Black, A.R., 1993, Predator-induced phenotypic plasticity in *Daphnia pulex*: life history and morphological responses to *Notonecta* and *Chaoborus*, *Limnology and Oceanography*, 38(5), pp. 986-96.

Boersma, M., Spaak, P. & De Meester, L., 1998, Predator-mediated plasticity in morphology, life history, and behavior of *Daphnia*: the uncoupling of responses, *The American Naturalist*, 152(2), pp. 237-48.

Bourdeau, P.E., 2010, An inducible morphological defence is a passive by-product of behaviour in a marine snail, *Proceedings. Biological sciences / The Royal Society*, 277(1680), pp. 455-62.

Buck, L.B., 1996, Information coding in the vertebrate olfactory system, *Annual review of Neuroscience*, 19(1), pp. 517-44.

Chapman, R.F. & Lee, J.C., 1991, Environmental effects on numbers of peripheral chemoreceptors on the antennae of a grasshopper, *Chemical Senses*, 16(6), pp. 607-16.

Colbourne J.K., Pfrender M.E., Gilbert D., Thomas W.K., Tucker A., Oakley T.H., Tokishita S., Aerts A., Arnold G.J., Basu M.K., Bauer D.J., Cáceres C.E., Carmel L., Casola C., Choi J.-H., Detter J.C., Dong Q., Dusheyko S., Eads B.D., Fröhlich T., Geiler-Samerotte K.A., Gerlach D., Hatcher P., Jogdeo S., Krijgsveld J., Kriventseva E.V., Kültz D., Laforsch C., Lindquist E., Lopez J., Manak J.R., Muller J., Pangilinan J., Patwardhan R.P. Pitluck S., Pritham E.J., Rechtsteiner A., Rho M., Rogozin I.B., Sakarya O., Salamov A., Schaack S., Shapiro H., Shiga Y., Skalitzky C., Smith Z., Souvorov A., Sung W., Tang Z., Tsuchiya D., Tu H., Vos H., Wang M., Wolf Y.I., Yamagata H., Yamada T., Ye Y., Shaw J.R., Andrews J., Crease T.J., Tang H., Lucas S.M., Robertson H.M., Bork P., Koonin E.V., Zdobnov E.V., Grigoriev I.V. Lynch M., Booreet J.L., 2011, The Ecoresponsive Genome of *Daphnia* pulex, *Science* 331(6017), pp. 555-561.

Crowl, T.A. & Covich, A.P., 1990, Predator-induced life-history shifts in a freshwater snail, *Science*, 247(4945), pp. 949-51.

Dodson, S.I., Tollrian, R. & Lampert, W., 1997, *Daphnia* swimming behavior during vertical migration, *Journal of Plankton Research*, 19(8), pp. 969-78.

Ebert D, 2005. Ecology, Epidemiology, and Evolution of Parasitism in *Daphnia* [Internet]. Bethesda (MD): National Library of Medicine (US),

National Center for Biotechnology Information. Available from: http://www.ncbi.nlm.nih.gov/entrez/query. fcgi?db=Books (Dec. 2011).

Ebert, D., 2011, A Genome for the Environment, *Science, 331(6017);* pp. 539.

Eisthen, H.L., 2002, Why are olfactory systems of different animals so similar? *Brain, Behavior and Evolution*, 59(5-6), pp. 273-93.

Elser, J.J., Fagan, W.F., Denno, R.F., Dobberfuhl, D.R., Folarin, A., Huberty, A., Interlandi, S., Kilham, S.S., McCauley, E. & Schulz, K.L., 2000, Nutritional constraints in terrestrial and freshwater food webs, *Nature*, 408(6812), pp. 578-80.

Emlen, D.J., Szafran, Q., Corley, L.S. & Dworkin, I., 2006, Insulin signaling and limb-patterning: candidate pathways for the origin and evolutionary diversification of beetle 'horns', *Heredity*, 97(3), pp. 179-91.

Fitzpatrick, M.J., Ben-Shahar, Y., Smid, H.M., Vet, L.E.M., Robinson, G.E. & Sokolowski, M.B., 2005, Candidate genes for behavioural ecology, *Trends in Ecology & Evolution*, 20(2), pp. 96-104.

Fritsch, M. & Richter, S., 2010, The formation of the nervous system during larval development in *Triops cancriformis* (Bosc) (Crustacea, Branchiopoda): An immunohistochemical survey, *Journal of Morphology*, 271(12), pp. 1457-81.

Fröhlich, T., Arnold, G.J., Fritsch, R., Mayr, T. & Laforsch, C., 2009, LC-MS/MS-based proteome profiling *in Daphnia pulex* and *Daphnia longicephala*: the *Daphnia pulex* genome database as a key for high throughput proteomics in *Daphnia, BMC genomics*, 10, p. 171.

Galizia, C.G. & Szyszka, P., 2008, Olfactory coding in the insect brain: molecular receptive ranges, spatial and temporal coding, *Entomologia Experimentalis et Applicata*, 128(1), pp. 81-92.

Gard, A.L., Lenz, P.H., Shaw, J.R. & Christie, A.E., 2009, Identification of putative peptide paracrines/hormones in the water flea *Daphnia pulex* (Crustacea; Branchiopoda; Cladocera) using transcriptomics and immunohistochemistry, *General and Comparative Endocrinology*, 160(3), pp. 271-87.

Gnatzy, W., Schmidt, M. & Römbke, J., 1984, Are the funnel-canal organs the "campaniform sensilla" of the shore crab *Carcinus maenas* (Crustacea, Decapoda)? *Zoomorphology*, 104(1), pp. 11-20.

Grant, J.W.G. & Bayly, I.A.E., 1981, Predator Induction of Crests in Morphs of the *Daphnia carinata* King Complex, *Limnology and Oceanography*, 26(2), pp. 201-18.

Hallberg, E., Johansson, K.U. & Elofsson, R., 1992, The aesthetasc concept: structural variations of putative olfactory receptor cell complexes in Crustacea, *Microscopy research and technique*, 22(4), pp. 325-35.

Harzsch, S., 2006, Neurophylogeny: Architecture of the nervous system and a fresh view on arthropod phyologeny, *Integrative and Comparative Biology*, 46(2), pp. 162-94.

Kiesecker, J.M., Chivers, D.P., Anderson, M. & Blaustein, A.R., 2002, Effect of predator diet on life history shifts of red-legged frogs, *Rana aurora*, *Journal of Chemical Ecology*, 28(5), pp. 1007-15.

Laforsch C. & Tollrian R., 2009, Cyclomorphosis and Phenotypic Changes. In: (ed) Gene E. Likens, *Encyclopedia of Inland Waters* Oxford: Elsevier Volume 3, pp. 643-650.

Lampert, W., 1989, The adaptive significance of diel vertical migration of zooplankton, *Functional Ecology*, 3(1), pp. 21-7.

Langerhans, R.B. & DeWitt, T.J., 2002, Plasticity constrained: over-generalized induction cues cause maladaptive phenotypes, *Evolutionary Ecology Research*, 4(6), pp. 857-70.

Le Trionnaire, G., Hardie, J., Jaubert-Possamai, S., Simon, J.C. & Tagu, D., 2008, Shifting from clonal to sexual reproduction in aphids: physiological and developmental aspects, *Biology of the Cell,* 100(8), pp. 441-51.

Marden, H., 2008, Quantitative and evolutionary biology of alternative splicing: how changing the mix of alternative transcripts affects phenotypic plasticity and reaction norms, *Heredity*, 100(2), pp. 111-20.

Mathis, A., Ferrari, M.C., Windel, N., Messier, F. & Chivers, D.P., 2008, Learning by embryos and the ghost of predation future, *Proceedings. Biological sciences*, 275(1651), pp. 2603-7.

Miyakawa, H., Imai, M., Sugimoto, N., Ishikawa, Y., Ishikawa, A., Ishigaki, H., Okada, Y., Miyazaki, S., Koshikawa, S., Cornette, R. & Miura, T., 2010, Gene up-regulation in response to predator kairomones in the water flea, Daphnia pulex, *BMC Developmental Biology*, 10, p. 45.

Moczek, A.P. & Rose, D.J., 2009, Differential recruitment of limb patterning genes during development and diversification of beetle horns, *Proceedings of the National Academy of Sciences*, 106(22), pp. 8992-7.

Montarolo, P.G., Goelet, P., Castellucci, V.F., Morgan, J., Kandel, E.R. & Schacher, S., 1986, A critical period for macromolecular synthesis in long-term heterosynaptic facilitation in *Aplysia*, *Science*, 234(4781), p. 1249.

Mougi, A. & Kishida, O., 2009, Reciprocal phenotypic plasticity can lead to stable predator-prey interaction, *The Journal of Animal Ecology*, 78(6), pp. 1172-81.

Nijhout, H.F., 2003, The control of growth, *Development*, 130(24), pp. 5863-7.

Oda, S., Kato, Y., Watanabe, H., Tatarazako, N. & Iguchi, T., 2011, Morphological changes in *Daphnia galeata* induced by a crustacean terpenoid hormone and its analog, *Environmental Toxicology and Chemistry,* 30(1), pp. 232-8.

Peñalva-Arana, D.C., Lynch, M. & Robertson, H.M., 2009, The chemoreceptor genes of the waterflea *Daphnia pulex*: many Grs but no Ors, *BMC Evolutionary Biology*, 9, p. 79.

Petrusek, A., Tollrian, R., Schwenk, K., Haas, A. & Laforsch, C., 2009, A "crown of thorns" is an inducible defense that protects *Daphnia* against an ancient predator, *Proceedings of the National Academy of Sciences of the United States of America*, 106(7), pp. 2248-52.

Pijanowska, J., 1992, Anti-predator Defence in Three *Daphnia* Species, *Internationale Revue der gesamten Hydrobiologie und Hydrographie*, 77(1), pp. 153-63.

Pijanowska, J. & Kowalczewski, A., 1997, Predators can induce swarming behaviour and locomotory responses in *Daphnia, Freshwater Biology*, 37(3), pp. 649-56.

Pijanowska, J., Dawidowicz, P. & Weider, L.J., 2006, Predator-induced escape response in Daphnia, *Archiv f_ü r Hydrobiologie, 167*, 1(4), pp. 77-87.

Price, T.D., Qvarnström, A. & Irwin, D.E., 2003, The role of phenotypic plasticity in driving genetic evolution, *Proceedings. Biological sciences*, 270(1523), pp. 1433-40.

Prudic, K.L., Jeon, C., Cao, H. & Monteiro, A., 2011, Developmental plasticity in sexual roles of butterfly species drives mutual sexual ornamentation, *Science*, 331(6013), pp. 73-5.

Rabus, M. & Laforsch, C., 2011, Growing large and bulky in the presence of the enemy: *Daphnia magna* gradually switches the mode of inducible morphological defences, *Functional Ecology*, 25(5), pp. 1137-43.

Riessen, H.P., 2012, Costs of predator-induced morphological defences in *Daphnia, Freshwater Biology*.

Sangster, T.A. & Queitsch, C., 2005, The HSP90 chaperone complex, an emerging force in plant development and phenotypic plasticity, *Current opinion in plant biology*, 8(1), pp. 86-92

Schmidt, M. & Ache, B.W., 1992, Antennular projections to the midbrain of the spiny. II. Sensory innervation of the olfactory lobe, *The Journal of Comparative Neurology*, 318(3), pp. 291-303.

Shingleton, A.W., Das, J., Vinicius, L. & Stern, D.L., 2005, The temporal requirements for insulin signaling during development in *Drosophila*, *PLoS Biology*, 3(9), p. e289.

Stibor, H., 1992, Predator induced life-history shifts in a freshwater cladoceran, *Oecologia*, 92(2), pp. 162-5.

Tautz, D., 2011, Not just another genome, *BMC biology*, 9(1), p. 8.

Tollrian, R., 1993, Neckteeth formation in *Daphnia pulex* as an example of continuous phenotypic plasticity: morphological effects of *Chaoborus* kairomone concentration and their quantification, *Journal of Plankton Research*, 15 (11)(11), pp. 1309-18.

Tollrian, R., 1994, Fish-kairomone induced morphological changes in *Daphnia lumholtzi* (Sars), *Archiv fur Hydrobiologie*, 130, pp. 69-69.

Tollrian, R. & Harvell, C.D., 1999, *The ecology and evolution of inducible defenses*, Princeton University Press, Princeton, New Jersey.

True, H.L., Berlin, I. & Lindquist, S.L., 2004, Epigenetic regulation of translation reveals hidden genetic variation to produce complex traits, *Nature*, 431(7005), pp. 184-7

Verschoor, A.M., Vos, M. & Van Der Stap, I., 2004, Inducible defences prevent strong population fluctuations in bi-and tritrophic food chains, *Ecology Letters*, 7(12), pp. 1143-8.

Vosshall, L.B., 2000, Olfaction in *Drosophila*, *Current Opinion in Neurobiology*, 10(4), pp. 498-503.

Weiss, L.C. 2012, The Taste of Predation and the Defences of Prey, in Oxford University Press, New York, pp. 111-26.

Weiss, L.C., Kruppert, S., Laforsch, C. & Tollrian, R., 2012a, *Chaoborus* and *Gasterosteus* Anti-Predator Responses in *Daphnia pulex* Are Mediated by Independent Cholinergic and Gabaergic Neuronal Signals, *PloS One*, 7(5), p. e36879.

Weiss, L.C., Tollrian, R., Herbert, Z. & Laforsch, C., 2012b, Morphology of the *Daphnia* nervous system: A comparative study on *Daphnia pulex*, *Daphnia lumholtzi*, and *Daphnia longicephala*, *Journal of Morphology*. 10.1002/jmor.20068

West-Eberhard, M.J., 2003, *Developmental plasticity and evolution*, Oxford University Press, USA.

Chapter 4

PHENOTYPIC PLASTICITY OF PLANTS IN RESPONSE TO ENVIRONMENTAL CHANGE

Mhemmed Gandour, Abdelly Chedly and Wael Taamalli
Laboratory of Extremophiles Plants, Center of Biotechnology
of Borj-Cédria, Tunisia

ABSTRACT

Plants can adjust their phenotype in response to changing environmental conditions through developmental plasticity. The study of phenotypic plasticity is naturally interdisciplinary and encompasses aspects of behaviour, development, ecology, evolution, genetics, genomics, and multiple physiological systems at various levels of biological organization. In this article, we provide a brief description of phenotypic plasticity in plants; examine its potential adaptive significance; emphasize recent molecular approaches that provide novel insight into underlying mechanisms, and highlight examples of phenotypic plasticity from plants behaviour in response to various environmental conditions.

INTRODUCTION

Global climate is changing at an exceptional rate. The increase in temperature over the two past decades, in particular, has been the warmest on record (Solomon et al., 2007). Climate change is a tough problem for law in many ways. It is possibly having profound and diverse effects on organisms. In particular, organisms living at mid- to high latitudes in the Northern Hemisphere are predicted to be the most affected by climate warming, because temperatures have risen most rapidly there (+3.4°C in the North versus -1.6°C in the south during the four last decades (Dillon et al., 2010)). It imposes strong selective pressures on natural populations (Peters and Lovejoy, 1994) and affects the physiology and development of individual organisms by reducing their growth rate and causing genetic evolution (Chevin, 2012). Indeed, it has a major impact on the viability of species. Whether a particular population will survive or go extinct will depend on its potential to adapt to new environmental conditions (Pulido, 2007). After an environmental shift, a population may change its geographic range to follow its optimum environment in space. However, dispersal can be limited by physical barriers and by natural or anthropogenic habitat fragmentation (Skole and Tucker, 1993). A population with a restricted geographic range can persist either by receiving migrants from a well-adapted source population (Pulliam, 1988), or by evolutionary adaptation to the new environment (Gomulkiewicz and Holt, 1995; Orr and Unckless, 2008). In the both cases the phenomena is called phenotypic plasticity. Phenotypic plasticity has been defined as the morphological and behavioral responses of flora to maximize their chances of survival in different sets of environmental circumstances or the production of environmentally adapted phenotypes by specific genotype. It has long been proposed to play a major role in the origin and subsequent diversification of morphological and behavioral novelties. Thus, it can serve as a pre-adaptation to various environmental changes. Fundamental questions for evolutionary ecologists in a global change context are how plant species will respond to these new and complex environmental scenarios and what mechanisms will be involved in the process. Until new, the few studies on the thematic of how plants respond to the climate change cannot cover all adaptation mechanism process. Rather we use available data to briefly discuss a few examples that illustrate how morphologic and genomic data can advance our understanding of the process of adaptation and refer interested readers to a number of excellent recent review articles summarizing specific topics.

HOW CLIMATE CHANGE?

Climate change refers to an increase in average global temperatures. Most climate scientists agree the main causes of the current global warming trend are human expansion of the "greenhouse effect" and the natural events. On the other side, scientists classify causes of climate change on two groups (figure 1): external (from extraterrestrial systems) and internal (from ocean, atmosphere and land systems). As a result, global atmospheric concentrations of CO2, CH4 and N2O have increased markedly (figure 2) which in turn has caused an increase of temperature (figure 3).

In fact, a change of existing climate conditions caused by the continued emission of large amounts of carbon dioxide into the atmosphere from anthropogenic activities will lead to higher global mean temperatures which in turn can result in a weakening or even a complete shutdown of the thermohaline circulation (Manabe and Stouffer, 1993; Rahmstorf and Ganopolski, 1999; Schmittner and Stocker, 1999). The consequences of such a shutdown are manifold:

The oceanic uptake of carbon may be reduced (Schmittner and Stocker, 1999), the climate of Northern and Western Europe will be severely affected (Broecker, 1997), and there may be pronounced economic and societal impacts. So that, the ocean is the major driver of global climate. It redistributes large amounts of heat around the planet via global ocean currents – through regional scale upwelling and downwelling, and via a process called thermohaline circulation.

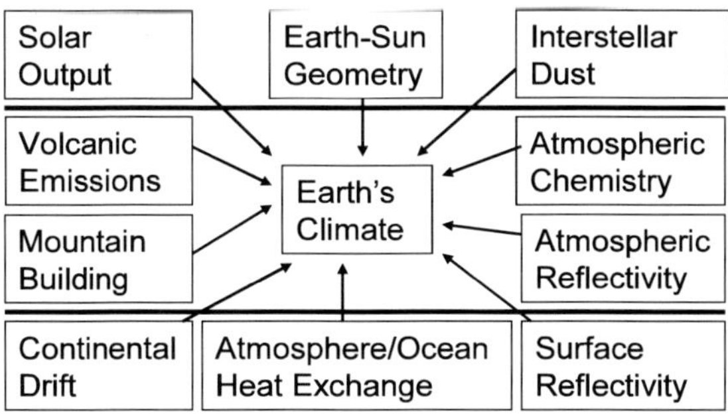

Figure 1. Factors that influence the Earth's climate (http://www. Physicalgeography .net/fundamentals/7y.html).

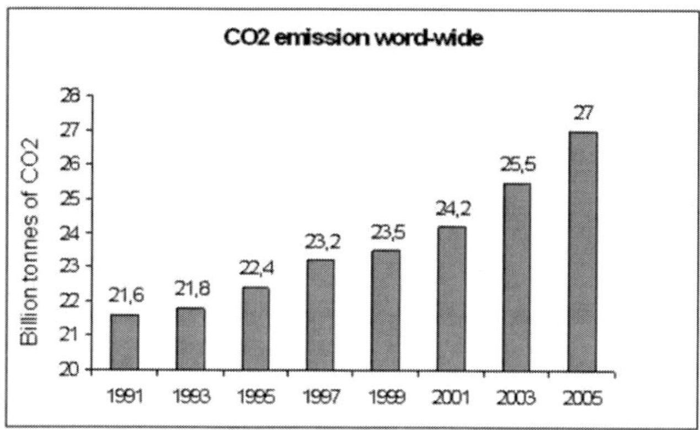

CO_2-emissions world-wide by year (data from wri.org).

Figure 2.

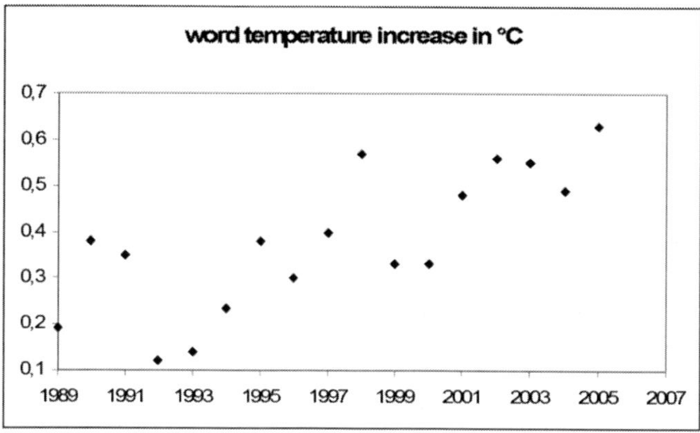

Increase of global average temperature for the last 20 years (source: wri.org).

Figure 3.

VIGOUR AND BEHAVIOUR OF PLANT

When a species encounters novel environmental conditions some phenotypic characters may develop differently than in the ancestral environment. In fact, natural selection would usually be expected to favor genetic changes that restore the ancestral phenotype or compensate for the environmental change in other ways (Conover and Schultz, 1995). Plastic responses span a broad variety of functional traits, as different components of global change affect different traits.

Particularly, it is well known that the leaf, compared with the other vegetative organs, is the best and fastest to react to changes in the environment. Consequently, the influence of different ecological factors on plant organisms is best reflected on leaf morphological and anatomical structure. For *Cakile* species, leaf shape is variable, becoming more pinnatifid towards the south and east of Europe (Ball, 1964) and towards the south of Tunisia (Gandour et al., 2008); this parallels a cline of increasing leaf dissection and may reflect selection for favourable thermal balance, with less transpiration, under higher irradiance during hotter, drier summers. Leaves produced early, or under adverse environmental conditions, tend to remain less pinnatifid. Although the leaves are succulent, they have thin cuticles and their epidermal cells are not regular in size, with a small fraction of slightly projecting cells 4–5 times the size of the others (Wright, 1927). There are more stomata on the abaxial (96 /mm^2) than on the adaxial (64 /mm^2) surface. Species adapted to specific climatic conditions must have adapted some structural adaptations (Nawazish et al., 2006) and leaf anatomical feature are the representative of such environmental adaptations.

For example *Eucalyptus camaldulensis* plants from more arid locations have thick leaves and high oil gland density (James and Bell, 1995). One other important trait is the flowering time. In fact, it is a good example of a crucial trait that has been shown to be both under genetic control and plastic one. In harassed conditions, flowering occurs in very small plants with reduced seed set rate and seed size, and in extreme cases no seed is produced. On the basis of this character, it has been suggested that short-day and long-day species respond differently to climatic change (Marc and Gifford, 1984; Reekie et al., 1994). Accordingly, Leakey et al. (2006) reported that environmental change has no or a little effect on plant C4. In fact, some organisms follow distinct developmental pathways in response to environmental cues by changing carbon allocation patterns. Specific adjustments can give rise to morphological and anatomical differences. Concerning the whole plant and biomass production, growth is much reduced under conditions of extreme nutrient paucity or lack of space. Both empirical and modeling studies have indicated that atmospheric and climatic changes can alter the functioning of terrestrial ecosystems via changes in net primary productivity (Norby and Luo, 2004).

In fact, elevated [CO2], warming, and altered precipitation can all directly and indirectly influence aboveground biomass production in terrestrial ecosystems. Elevated [CO2] can stimulate plant growth through enhanced photosynthesis (Tolbert and Zelitch, 1983; Field et al., 1995) and reduce water use through stomatal closure (Owensby et al., 1993), generally resulting in greater net primary productivity (Dukes et al., 2005; Norby et al., 2005).

PLANT MIGRATION IN RESPONSE TO CLIMATE CHANGE

Evidence inferred from past climate changes tends to indicate that species are more likely to respond by migration rather than by adapting genetically (Huntley, 1991). This migration may be in the form of habitat expansion through the outlier individuals or the success of hitch-hiking. Indeed, climate is considered the primary factor controlling plant distribution (Woodward, 1987). Phenotypic plasticity during migration is the colonizing genotypes express a suboptimal phenotype in the new environment. Directional selection will favor the phenotypes closest to the optimum in the new environment, therefore facilitating adaptation to the new conditions. Parmesan and Yohe (2003) reported that the distribution of many plant species has already altered in response to climate change; some species have shown up to 6 km pole-ward migration each year over the past 16–132 years. In tropical zone, some scientists estimate that the rate of plant moving is 15 feet per day, or about a mile per year so that 17.6 kilometers per decade, which is three times the rate previously described. They assert also that the plants moving the most are in the areas with the highest levels of warming. Given the projected changes in climate over the next 100 years will most likely exceed the natural ability of populations to acclimate, adapt in place, or migrate (Rehfeldt et al., 2001; Iverson et al., 2004; Neilson et al., 2005; Aitken et al., 2008), proactive human mediated transportation of tree species populations to more favourable climatic habitats may be a management option (Ledig and Kitzmiller, 1992; Rehfeldt et al., 1999, 2001; Beaulieu and Rainville, 2005; Millar et al., 2007; Campbell et al., 2009).

MECHANISMS

Because natural plant populations are sessile organisms and cannot move from away stressful environment, they have to develop and coordinate diverse types of mechanisms to respond to the changing environment to sustain survival in the soil–plant–atmosphere continuum environment (Hong-bo,

2006). When we look inside the adaptation process, we can see that many mechanisms have been integrated to promote this adaptation. In addition to genotype which considered for a long time as the main contributor of phenotypic variation, many studies have confirmed recently that epigenotype has added another level of complexity to the system and have confirmed their contribution to phenotypic plasticity towards a rapid change of transcriptional profile by epigenetic regulators (Reusch and Wood, 2007). These regulators may induce different transcriptomic adaptations that determine the fundamental niches of populations (Reusch and Wood, 2007). Plant respiration is one of the key processes in terms of an understanding of plant growth and functioning in a future climate. Some tolerant plants would response mainly via improving osmotic adjustment ability and increasing cell wall elasticity to maintain tissue turgidity (Morgan et al., 1984) others via altering metabolic path for life survives under sever stress (Bartoli et al., 1999 and Peñuelas et al., 2004). One of the major modifications is the increase of metabolites (e.g. increased antioxidant metabolism). In fact, the metabolic rates increased most quickly in the tropics and north temperate zones, and less so in the Arctic. Accordingly Dillon et al. (2010) reported that, during the past three decades, warming has had its biggest absolute impacts on metabolic rates in tropical and north temperate zones. At the epigenetic processes, gene expression can be altered by many process including DNA methylation, histone modification and transposable element activation, changes to the population of small RNAs can also alter gene expression and by turn can modify plasticity (Chinnusamy et al. 2009).

GENE DUPLICATION AS A MECHANISM OF GENOMIC ADAPTATION TO A CHANGING ENVIRONMENT

It is clear that some gene duplications have been fixed in the course of evolution by positive selection (Velkov, 1982; Shaner et al., 2012). Recent studies demonstrate that these gene duplications have an adaptive short-term role. Furthermore, a consistent polyploidy of the yeast strains was observed that may play a role in the adaptation (Dhar et al., 2011; Gerstein et al., 2006). In fact, the adaptive hypothesis holds that function is a quantitative measure of gene action that can be influenced by the amount of gene product in the cell, which in turn can be influenced by the gene copy number (Hahn, 2009; Innan and Kondrashov, 2010). For example, adaptation to stress by high salt content may occur through gene duplication. In fact, polyploidy has been linked to resistance to high salt concentrations in citrus (Saleh et al., 2008) and sorghum (Ceccarelli et al., 2006) suggesting that polyploidy may be a general physiological adaptive response to osmotic stress (Dhar et al., 2011).

GENETIC

There is general acceptance that high levels of genetic variation within natural populations improve the potential to survive and adapt to novel biotic and abiotic environmental changes including the tolerance of climatic change (Jump, A.S. et al., 2009).

In fact, genetic variation in genes and transcription factors could help plant populations to adapt to rapid changes. Furthermore, new gene combinations, arising from genetic recombination, can enhance the plant's tolerance or resistance to environmental stresses. That genotype may contribute to phenotypic variation often responding differently to environmental variation (Richards et al., 2010) and consequently can both provide a buffer against rapid climate changes and assist rapid adaptation (Lande, 2009; Chevin et al., 2010). Recent studies have investigated the importance of gene-by-environment interactions, epistasis and adaptive genes during plant stress (Kumar, 1999; Pinto et al., 2010). The variation on a locus may confer stress tolerance, but reduced variation on a locus may be the result of strong directional selection that fixed a single allelic variant that conferred stress tolerance.

In natural environments, the stress condition may direct the selection of trait combinations that confer higher fitness, decreasing the genetic variation in the generations.

Genetic diversity is an important prerequisite for adapting to new environmental conditions. Certain evidence suggests that strong selection like that induced by a climatic change may create very rapid genetic differentiation within plant populations (Theurillat et al., 1998). The present distribution of many regional endemic species has been explained by a genetic impoverishment which prevented a greater expansion (Niklefeld, 1972). According to recent results, observed genetic variation along an ecological gradient can result from selection, e.g., in *Cakile maritima* along a latitudinal factor (Gandour et al., 2008). In addition, the genetic diversity of a species itself can vary between distinct populations in separated geographical areas of its range.

However, the rate of environmental change is faster than the rate of new gene combinations (Peng and Zhang, 2009). In this context, genetic changes in plant genomes have to cope with the different types of stress to which they are exposed.

HYBRIDIZATION AND ADAPTATION

Many genera in temperate and boreal regions contain species pairs capable of interspecific hybridization, including spruce (*Picea*), pine (*Pinus*), poplar (*Populus*), and oak (*Quercus*). Such phenomenon makes species more adaptive to new environmental conditions by providing novel alleles for adaptations (Morjan and Rieseberg, 2004). Hybridization followed by backcrossing or further introgression and selection can lead to transgressive segregation, whereby some individuals can have phenotypes outside of the range of parental species due to complementary allelic effects (Lexer et al., 2004). This may offer a rapid evolutionary path to adaptation to novel environments (Rieseberg et al., 2003).

REFERENCES

Aitken, S. N.; Yeaman, S.; Holliday, J.A.; Wang, T. and Curtis-McLane S. (2008) Adaptation, migration or extirpation: climate change outcomes for tree populations. *Evol. Appl.,* 1: 95-111.

Ball B.W. (1964) *Cakile* Miller. In Tutin T.G., Heywood V.H., Burges N.A., Valentine D.H., Walters S.M. and Webb D.A. (Eds), "Flora Europaea", vol. 1, p 343. Cambridge University Press, Cambridge.

Bartoli, C.G.; Simontacchi, M.; Tambussi, E.; Beltrano, J.; Montaldi, E. and Puntarulo, S. (1999) Drought and watering-dependent oxidative stress: effect on antioxidant content in Triticum aestivum L. leaves. *J. Exp. Bot.,* 50: 375-85.

Beaulieu, J. and Rainville, A. (2005) Adaptation to climate change: Genetic variation is both a short- and long-term solution. *For. Chron.,* 81: 704-709.

Campbell, E.; Saunders, S.C.; Coates, D.; Meidinger, D.; MacKinnon, A.; O'Neill, G.; MacKillop, D. and DeLong, C. (2009) Ecological resilience and complexity: a theoretical framework for understanding and managing

British Columbia's forest ecosystems in a changing climate. BC Min. For. Range, For. Sci. Prog., Victoria, BC. *Tech. Rep.,* 055. 36p.

Ceccarelli, M.; Santantonio, E.; Marmottini, F.; Amzallag, G.N. and Cionini, P.G. (2006) Chromosome endoreduplication as a factor of salt adaptation in Sorghum bicolor. *Protoplasma,* 227: 113–118.

Chevin, L-M.; Lande, R. and Mace, G.M. (2010) Adaptation, plasticity, and extinction in a changing environment: towards a predictive theory. *PLoS Biol.,* 8: e1000357.

Chinnusamy, V. and Zhu, J.K. (2009) Epigenetic regulation of stress responses in plants. *Curr. Opin. Plant Biol.,* 12, 133–139.

Conover, D.O. and Schultz E.T. (1995) Phenotypic similarity and the evolutionary significance of counter gradient variation. *Trends in Ecology and Evolution,* 10: 248–252.

Dhar, R.; Sägesser, R.; Weikert, C.; Yuan, J. and Wagner A. (2011) Adaptation of Saccharomyces cerevisiae to saline stress through laboratory evolution. *J. Evol. Biol.,* 24: 1135–1153.

Dillon, M.E.; Wang G. and Huey R.B. (2010) Global metabolic impacts of recent climate warming. *Nature,* 467: 704-707.

Dukes, J.S.; Chiariello, N.R.; Cleland, E.E.; Moore, L.A.; Shaw, M.R.; Thayer, S.; Tobeck, T.; Mooney, H.A. and Field C.B (2005) Responses of grassland production to single and multiple global environmental changes. *PLoS Biology,* 3: 1829–1837.

Field, C.B.; Jackson, R.B. and Mooney H.A. (1995) Stomatal responses to increased CO2 - implications from the plant to the global-scale. Plant, *Cell and Environment,* 18: 1214–1225.

Gandour M., Hessini K. and Abdelly C. 2008. Long distance seed dispersal: role of sea currents in determining population structure of Cakile maritima. *Genetics research*, 90, 167-178.

Gerstein, A.C.; Chun, H.J.; Grant, A. and Otto S.P. (2006) Genomic convergence toward diploidy in *Saccharomyces cerevisiae. PLoS Genet.,* 2: e145.

Gomulkiewicz, R. and Holt, R.D. (1995) when does natural selection prevent extinction? *Evolution,* 49: 201–207.

Hahn, M.W. (2009) Distinguishing among evolutionary models for the maintenance of gene duplicates. *J. Hered.,* 100: 605–617.

Hong-bo, S.; Li-ye, C.; Chang-xing, Z.; Qing-jie, G.; Xian-an, L. and Jean-Marcel, R. (2006) Plant gene regulatory network system under abiotic stress, *Acta Biol. Szegediensis,* 50: 1–9.

Huntley, B. (1991) 'How Plants Respond to Climate Change: Migration Rates, Individualism and the consequences for Plant Communities', *Ann. Botany*, 67: 15–22.

Innan, H. and Kondrashov, F. (2010) the evolution of gene duplications: classifying and distinguishing between models. *Nat. Rev. Genet.*, 11: 97–108.

Iverson, L.R.; Schwartz, M.W. and Prasad, A.M. (2004) How fast and far might tree species migrate in the eastern United States due to climate change? *Global Ecol. Biogeogr.*, 13: 209-219.

James, S.A. and Bell. D.T. (1995) Morphology and anatomy of leaves of Eucalyptus camaldulensis clones: Variation between geographically separated locations. *Aust. J. Bot.*, 43: 415-433.

Jump A.S.; Marchant, R. and Peñuelas, J. (2009) Environmental change and the option value of genetic diversity. *Trends Plant Sci.*, 14: 51–58.

Kumar, L.S. (1999) DNA markers in plant improvement: an overview, *Biotechnol. Adv.*, 17: 143–182.

Lande, R. (2009) Adaptation to an extraordinary environment by evolution of phenotypic plasticity and genetic assimilation. *J. Evol. Biol.*, 22: 1435–1446.

Leakey, A.; Uribelarrea, M.; Ainsworth, E.; Naidu, S.; Rogers, A.; Ort, D. and Long, S.P. (2006) Photosynthesis, productivity, and yield of maize are not affected by open-air elevation of CO2 concentration in the absence of drought. *Plant Physiology*, 140: 779-790.

Ledig, F.T. and Kitzmiller, J.H. (1992) Genetic strategies for reforestation in the face of climate change. *For. Ecol. Manage*, 50: 153-169.

Lexer, C.; Heinze, B.; Alia, R. and Rieseberg, L.H. (2004) Hybrid zones as a tool for identifying adaptive genetic variation in outbreeding forest trees: lessons from wild annual sunflowers (Helianthus spp.). *Forest Ecology and Management*, 197: 49–64.

Luis-Miguel Chevin (2012) Genetic constraints on adaptation to a changing environment. *Evolution*, 1558-5646.

Manabe, S. and Stouffer, R.J. (1993): Century-scale effects of increased atmospheric CO2 on the ocean-atmosphere system, Nature, 364: 215-218.

Marc, J. and Gifford, R.M. (1984) Floral initiation in wheat, sunflower, and sorghum under carbon dioxide enrichment. *Canadian Journal of Botany*, 62: 9-14.

Millar, C.I.; Stepehenson, N.J. and Stephens S.L. (2007) Climate change and forests of the future: managing in the face of uncertainty. *Ecol. Appl.*, 17: 2145-2151.

Morgan, J.M. (1984) Osmoregulation and water stress in higher plants. *Ann. Rev. Plant Physiol.*, 35: 299-319.

Morjan, C.L. and Rieseberg L.H. (2004) How species evolve collectively: implications of gene flow and selection for the spread of advantageous alleles. *Molecular Ecology,* 13: 1341–1356.

Nawazish, S.; Hameed, M. and Naurin, S. (2006) Leaf anatomical adaptations of Cenchrus ciliaris l. from the Salt Range, Pakistan against drought stress. *Pak. J. Bot.,* 38: 1723-1730.

Neilson, R.P.; Pitelka, L.F.; Solomon, A.M.; Nathan, R.; Midgley, G.F.; Fragaso, J.M.V.; Lischke, H. and Thompson, K. (2005) Forecasting regional to global plant migration in response to climate change. *Bioscience*, 55: 749-759.

Niklfeld, H. (1972) 'Der niederösterreichische Alpenostrand – ein Glazial Refugium montaner Pflanzensippen', Jahrb. Vereins Schutze Alpenpfl. *Tiere,* 37: 42–94.

Norby, R.J.; DeLucia, E.H.; Gielen B., Calfapietra, C.; Giardina, C.P.; King, J.S.; Ledford, J.; McCarthy, H.R.; Moore, D.J.P.; Ceulemans, R.; De Angelis, P.; Finzi, A.C.; Karnosky, D.F.; Kubiske, M.E.; Lukacm, M.; Pregitzer, K.S.; Scarascia-Mugnozza, G.E.; Schlesinger, W.H. and Oren, R. (2005) Forest response to elevated CO2 is conserved across a broad range of productivity. *Proceedings of the National Academy of Sciences of the United States of America,* 102: 18052–18056.

Norby, R.J. and Luo, Y. (2004) Evaluating ecosystem responses to rising atmospheric CO2 And global warming in a multifactor world. *New Phytologist,* 162: 281–293.

Orr, H.A. and Unckless R.L. (2008) Population extinction and the genetics of adaptation. *Am. Nat.,* 172: 160–169.

Owensby, C.E.; Coyne, P.I.; Ham, J.M.; Auen, L.M. and Knapp, A.K. (1993) Biomass production in a tallgrass prairie ecosystem exposed to ambient and elevated CO2. *Ecological Applications,* 3: 644–653.

Parmesan, C. and Yohe, G. (2003) A globally coherent fingerprint of climate change impacts across natural systems. *Nature,* 421: 37–42.

Peng, H. and Zhang, J. (2009) Plant genomic DNA methylation in response to stresses: potential applications and challenges in plant breeding, *Prog. Nat. Sci.,* 19: 1037–1045.

Peñuelas, J.; Munné-Bosch, S.; Llusià, J. and Filella, I. (2004) Leaf reflectance and photo- and antioxidant protection in field-grown summer-stressed Phillyrea angustifolia. Optical signals of oxidative stress? *New Phytol.,* 162: 115-24.

Peters, R.L. and Lovejoy, T.E. (1994) Global warming and biological diversity. New Haven CT: Yale University Press.

Pinto, R.S.; Reynolds, M.P.; Mathews, K.L.; McIntyre, C.L.; Olivares-Villegas, J.J. and Chapman, S.C. (2010) Heat and drought adaptive QTL in a wheat population designed to minimize confounding agronomic effects, *Theor. Appl. Genet.*, 121: 1001–1021.

Pulido, F. (2007) Phenotypic changes in spring arrival: evolution, phenotypic plasticity, effects of weather and condition. *Clim. Res.*, 5-23.

Pulliam, H.R. (1988) Sources, sinks, and population regulation. *Am. Nat.*, 132: 652–661.

Rahmstorf, S. and Ganopolski, A. (1999): Long-term global warming scenarios computed with an efficient coupled climate model, *Climatic Change*, 43: 353-367.

Reekie, J.; Hicklenton, P. and Reekie, E. (1997) The interactive effects of carbon dioxide enrichment and daylength on growth and development in Petunia hybrida. *Annals of Botany,* 80: 57-64.

Rehfeldt, G.E.; Wykoff, W.R. and Ying, C.C. (2001) Physiological plasticity, evolution and impacts of a changing climate on Pinus contorta. *Clim. Change*, 50: 355-376.

Rehfeldt, G.E.; Wykoff, W.R. and Ying C.C. (2001) Physiological plasticity, evolution, and impacts of a changing climate on Pinus contorta. *Clim. Change*, 50: 355-376.

Rehfeldt, G.E.; Ying, C.C.; Spittlehouse, D.L. and Hamilton, D.A. (1999) Genetic response to climate in Pinus contorta: nice breadth, climate change, and reforestation. *Ecol. Monogr.*, 69: 375-407.

Reusch, T.B.H. and Wood, T.E. (2007) *Molecular ecology of global change,* *Mol. Ecol.*, 16: 3973–3992.

Richards, C.L.; Bossdorf, O. and Pigliucci, M. (2010) What role does heritable epigenetic variation play in phenotypic evolution? *Bioscience,* 60: 232–237.

Rieseberg, L.H.; Raymond, O.; Rosenthal, D.M.; Lai, Z.; Livingstone, K.; Nakazato, T.; Durphy, J.L.; Schwarzbach, A.E.; Donovan, L.A. and Lexer C. (2003). Major ecological transitions in wild sunflowers facilitated by hybridization. *Science,* 301: 1211–1216.

Saleh, B.; Allario, T.; Dambier, D.; Ollitrault, P. and Morillon, R. (2008) Tetraploid citrus rootstocks are more tolerant to salt stress than diploid. C. R. Biol., 331: 703–710.

Schmittner, A. and Stocker, T.F. (1999): The Stability of the Thermohaline Circulation in Global Warming Experiments, *Journal of Climate*, 12: 1117-1133.

Shaner, D.L.; Lindenmeyer, R.B. and Ostlie, M.H. (2012) What have the mechanisms of resistance to glyphosate taught us? *Pest Manag. Sci.*, 68: 3–9.

Skole, D. and Tucker, C.J. (1993) Tropical deforestation and habitat fragmentation in the Amazon: satellite data from 1978 to 1988. *Science*, 260: 1905–1910.

Solomon, S.; Qin, D.; Manning M., et al., (2007) "Technical Summary" in Climate Change 2007: Physical Science Basis. Contribution of Working Group I to the Fourth Assessment Report of the Intergovernmental Panel on Climate Change [Solomon, S., D. Qin, M. Manning, Z. Chen, M. Marquis, K.B. Averyt, M. Tignor and H.L. Miller (eds.)]. Cambridge University Press, Cambridge, United Kingdom and New York, NY, USA.

Theurillat, J.P.; Felber, F.; Geissler, P.; Gobat, J.M.; Fierz, M.; Fischlin, A.; Küpfer, P.; Schlüssel, A.; Velutti, C. and Zhao G.F. (1998) 'Sensitivity of Plant and Soils Ecosystems of the Alps to Climate Change', in Cebon, P.; Dahinden, U.; Davies, H.C.; Imboden, D. and Jaeger, C.C. (eds.), 'Views from the Alps: Regional Perspectives on Climate Change', MIT Press, Cambridge, MA, pp. 225–308.

Tolbert, N.E. and Zelitch, I. (1983) Carbon metabolism. In: CO2 and Plants: The Response of Plants to Rising Levels of Atmospheric Carbon Dioxide (ed Lemon ER), pp. 21–64. Westview, Boulder, CO.

Velkov, V.V. (1982) Gene amplification in prokaryotic and eukaryotic systems. *Genetika*, 18: 529–543.

Woodward, F.I. (1987) Climate and plant distribution. Cambridge University Press, New York, New York, USA.

Wright, J. (1927) Notes on strand plants. II. Cakile maritima Scop. *Transactions and Proceedings of the Botanical Society of Edinburgh*, 29: 389-401.

http://www.physicalgeography.net/fundamentals/7y.html.www.wri.org.

INDEX

D

E

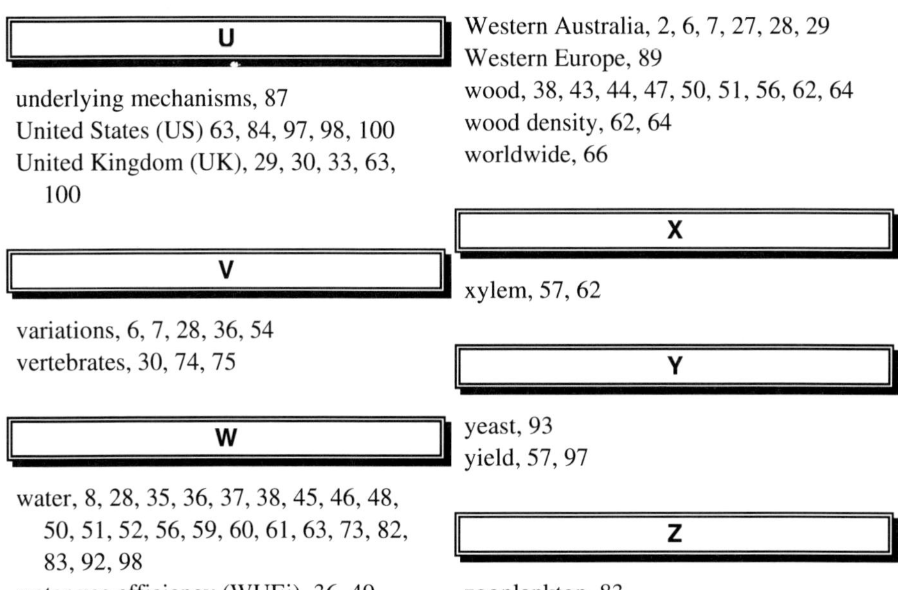